Guidelines for Food Waste Prevention and Reduction: Insights from the Fruit and Vegetable Sector

Iulia Ajtai • Dacinia Crina Petrescu
Ruxandra Malina Petrescu-Mag
Editors

Guidelines for Food Waste Prevention and Reduction: Insights from the Fruit and Vegetable Sector

Springer

Editors
Iulia Ajtai
Faculty of Environmental Science
and Engineering
Babes-Bolyai University
Cluj-Napoca, Cluj, Romania

Dacinia Crina Petrescu
Faculty of Business
Babes-Bolyai University
Cluj-Napoca, Cluj, Romania

Ruxandra Malina Petrescu-Mag
Faculty of Environmental Science
and Engineering
Babes-Bolyai University
Cluj-Napoca, Cluj, Romania

ISBN 978-3-032-14505-5 ISBN 978-3-032-14506-2 (eBook)
https://doi.org/10.1007/978-3-032-14506-2

This Springer imprint is published by the registered company Springer Nature Switzerland AG
The registered company address is: Gewerbestrasse 11, 6330 Cham, Switzerland

CRediT Authorship Contribution Statement

Iulia Ajtai: Conceptualization, Data collection (Interviews), Formal analysis, Data curation, Writing—original draft, review and editing, Transcripts, Visualization; Dacinia Crina Petrescu: Conceptualization, Methodology, Data collection (Interviews), Formal analysis, Data curation, Writing—original draft, review and editing, Project administration, Supervision, Funding; Ruxandra Malina Petrescu-Mag: Conceptualization, Methodology, Data collection (Interviews), Writing: review and editing; Ioana Cristina Piştea, Carmen Andreea Roba: Data collection (Interviews), Transcripts; Oana Adriana Gică, Ancuţa Ţenter, Valentin Toader, Adina Letiţia Negruşa, Daniel Măluţan, Nicoleta Bican-Brişan: Data collection (Interviews).

Infographics and figures: Iulia Ajtai

Image and template sources for infographics and figures: canva.com and freepik.com (based on a subscription account)

Acknowledgment

The authors are grateful to the participants in the project's interviews for their time, valuable opinions, and insights. They are an inspiration for the project team through their commitment to fighting food waste.

The "Team Up for a Circular Economy: A Collaborative Approach between Farmers, Retailers, and Consumers for Less Food Wastage" (hereinafter "Team Up for Less Food Waste") project was funded by a grant from the Ministry of Research, Innovation and Digitization, CNCS – UEFISCDI, Romania, project number PN-III-P4-PCE-2021-0538, within PNCDI III. The project was implemented by a team of researchers from Babes-Bolyai University, Cluj-Napoca, Romania, at the Faculty of Business between 2022 and 2024. Project director (Principal Investigator): Prof. Dr. Dacinia Crina Petrescu.

Editors:

Iulia Ajtai; Dacinia Crina Petrescu; Ruxandra Malina Petrescu-Mag

Authors:

Iulia Ajtai; Dacinia Crina Petrescu; Ruxandra Malina Petrescu-Mag; Ioana Cristina Piştea; Carmen Andreea Roba; Oana Adriana Gică; Ancuţa Ţenter; Valentin Toader; Adina Letiţia Negruşa; Daniel Măluţan; Nicoleta Bican-Brişan

Corresponding editor and author: Dacinia Crina Petrescu. Email: crina.petrescu@ubbcluj.ro

Contents

List of Figures

List of Infographics

Chapter 1
How This Book Emerged: The "Team Up for Less Food Waste" Project

Dacinia Crina Petrescu

This book contains the results and lessons learned during the research project "Team up for a circular economy: A collaborative approach between farmers, retailers, and consumers for less food wastage" (hereinafter, "Team Up for Less Food Waste"). The project was funded by a grant from the Ministry of Research, Innovation and Digitization, CNCS—UEFISCDI, Romania, project number PN-III-P4-PCE-2021-0538, within PNCDI III. The project was implemented by a team of researchers from Babes-Bolyai University, Cluj-Napoca, Romania, at the Faculty of Business between 2022 and 2024. This brief presentation of the premises and purpose of the "Team Up for Less Food Waste" project aims to explain why the project was necessary and what its ambitions are.

Transition to sustainable food systems is part of the New Green Deal and one of the 27 objectives of the EU's Farm-to-Fork Strategy, whose main aim is to halve per capita food waste by 2030 (FAO, 2022). The revised EU waste legislation (adopted in May 2018) demands that all EU countries take action to reduce food waste at each stage of the food supply chain, monitor, and report on food waste levels. Food waste is a pervasive issue found at every stage of the food supply chain (Göbel et al., 2015; Read et al., 2020), highlighting the necessity for collaborative efforts across the entire chain to effectively prevent and reduce it. Fruit and vegetables (FV) (considered together) are the most wasted foods (in volume) among various food categories (FAO, 2015).

Romania is a significant producer of FV within the EU, ranking in sixth place for fruits, berries, and nuts (with 1,436,190 t in 2022, harvested production in EU standard humidity) and in ninth place for fresh vegetables (with 1,433,279 t in 2022,

D. C. Petrescu (✉)
Faculty of Business, Babes-Bolyai University, Cluj-Napoca, Romania
e-mail: crina.petrescu@ubbcluj.ro

I. Ajtai et al. (eds.), *Guidelines for Food Waste Prevention and Reduction: Insights from the Fruit and Vegetable Sector*, https://doi.org/10.1007/978-3-032-14506-2_1

harvested production in EU standard humidity) (EUROSTAT, 2024a, 2024b, 2024e). Regarding food waste, Romania held the ninth highest level of food waste in the EU in 2022 (European Environment Agency, 2023). At the same time, food products represented the largest share of household final consumption expenditure by consumption purpose (accounting for 23.7% in 2022, the highest proportion among the 27 EU countries) (EUROSTAT, 2024c, 2024d). This is particularly relevant given that households are the biggest generators of food waste among various sectors (EUROSTAT, 2024e). In the same year, the GDP per capita in Romania was the lowest in the EU (The World Bank, 2024). Despite this worrying situation, there is currently a dearth of available data concerning details about FV waste in Romania (Archip et al., 2023).

In this context, the "Team Up for Less Food Waste" project emerged with the intention of stimulating collaboration between food value chain members as a means to prevent and reduce food waste within the circular economy paradigm. The project aimed to contribute to food waste prevention and reduction efforts by specifically targeting unprocessed FV waste and the stakeholders of the FV food value chain. The project's objectives discussed in this book were (i) to reveal the meaning of FV waste according to stakeholders' understanding, (ii) to cast light on the causes of FV waste as perceived by stakeholders, and (iii) to identify stakeholders' preferred solutions for preventing and reducing FV waste.

According to the Waste Hierarchy of the EU Waste Framework Directive, the best solution for the management of food by-products and food waste is prevention, followed by its reduction associated with various valorization options (preparing for reuse, recycling, and other material recovery), while disposal is the least preferred option (European Commission, n.d.; Obersteiner & Scherhaufer, 2021). The "Team Up for Less Food Waste" project adhered to the principles outlined in the waste

Fig. 1.1 The five-step "waste hierarchy" of the EU waste management (Source: European Commission, n.d.)

hierarchy, which indicates the order of preference for managing and disposing of waste (Fig. 1.1).

The project "Team Up for Less Food Waste" adopted a mixed-methods approach, combining interviews, surveys, and experiments in the farmers' market. This book presents the results of the qualitative research stage that included over 40 interviews with members of the FV food value chain. In this way, we aim to facilitate collaboration among members of the FV food value chain by shedding light on their perspectives regarding food waste, including food waste definition, causes, and solutions for preventing and reducing FV waste.

Chapter 2
Food Waste: Definition, Impacts, and Key Statistics

Iulia Ajtai, Dacinia Crina Petrescu, Ruxandra Malina Petrescu-Mag, Ioana Cristina Piştea, Carmen Andreea Roba, Oana Adriana Gică, Ancuţa Ţenter, Valentin Toader, Adina Letiţia Negruşa, Daniel Măluţan, and Nicoleta Bican-Brişan

2.1 What Is and What Is Not Fruit and Vegetable Waste

Our understanding of what constitutes food waste and what does not shapes all actions and perspectives related to it. The definition of food waste will guide reduction strategies, measurement methods, and good practices and influence interest in and commitment to them. Comparative analysis of food waste levels between countries or sectors also requires a clear definition and understanding of the food waste concept, as well as comparable measurement methods. Two tasks emerge concerning the meaning of food waste. One is to provide a clear definition, and the other is to ensure harmonization of meanings, such as between the legal definition and stakeholders' understanding of food waste, while allowing enough flexibility to adapt to particular contexts.

Pursuing the former task, this chapter presents the most common meanings assigned to food waste and associated concepts in the literature, as described by Petrescu-Mag et al. (2024). There are various words in the "food waste" family of concepts, such as loss, waste, and wastage, and their meaning differs in the literature and practice (Boiteau & Pingali, 2023). Additionally, there are differences in interpreting these words across various languages. For example, some authors consider intentionality and use "waste" for what is voluntary and "loss" for what is not, while

I. Ajtai · R. M. Petrescu-Mag · I. C. Piştea · C. A. Roba · A. Ţenter · N. Bican-Brişan
Faculty of Environmental Science and Engineering, Babes-Bolyai University, Cluj-Napoca, Romania

D. C. Petrescu (✉) · O. A. Gică · V. Toader · A. L. Negruşa · D. Măluţan
Faculty of Business, Babes-Bolyai University, Cluj-Napoca, Romania
e-mail: crina.petrescu@ubbcluj.ro

I. Ajtai et al. (eds.), *Guidelines for Food Waste Prevention and Reduction: Insights from the Fruit and Vegetable Sector*, https://doi.org/10.1007/978-3-032-14506-2_2

others take edibility into account and consider two categories of waste—“non-avoidable waste” (non-edible parts of food) and “avoidable waste” (edible food waste) (HLPE, 2014). Furthermore, food “loss” and “waste” are used distinctively according to the stage of the food supply chain where loss or waste takes place: “loss” is used for the beginning of the supply chain (i.e., from harvest to retailer but without including the retail level) and “waste” for the end of it (retail and consumer level) (Aragie et al., 2018; Gustavsson et al., 2011; Rezaei & Liu, 2017). The Food and Agriculture Organization (FAO, the United Nations) also makes this distinction between food loss and waste, defining food loss as a “decrease in edible food mass at the production, post-harvest, and processing stages of the food chain,” food waste as the “discard of edible foods at the retail and consumer levels,” and food wastage as the sum of both food loss and waste (FAO, 2013, 2023).

A well-known definition of food waste is the one provided by the EU FUSIONS (2016) project: “any food, and inedible parts of food, removed from the food supply chain to be recovered or disposed of (including composted, crops plowed in/not harvested, anaerobic digestion, bio-energy production, co-generation, incineration, disposal in sewer, landfill or discarded to sea).” It is important to highlight that FUSIONS considers that all supply chain members generate food waste. However, not all crops can be “food waste.” FUSIONS’ definition of food waste includes only the mature crops or those in the post-harvest stage (i.e., crops plowed into the soil, left unharvested, spoiled during manipulation or storage). FUSIONS does not include in the “food waste” category the unharvested or lost crops at the primary producer level that did not reach harvest maturity, nor the food that becomes animal feed or bio-based materials. The rationale behind this definition is that manufacturing animal feed or bio-based materials from food can be just as resource-efficient as producing food itself, potentially yielding economic benefits for the company involved in the production process (European Commission, 2021; FUSIONS, 2016).

Within the European Union (EU) legal framework, the “food waste” concept is defined in the revised Waste Framework Directive (Article 1, point 3, c) as “all food as defined in Article 2 of Regulation (EC) No. 178/2002 of the European Parliament and of the Council (*2) that has become waste” (European Parliament & Council of Ministers, 2018). To fully understand what the EU means by “food waste,” the definitions of “food” and “waste” should also be mentioned. Thus, “any substance or product, whether processed, partially processed or unprocessed, intended to be, or reasonably expected to be ingested by humans” is considered “food,” but “food” does not include feed and plants prior to harvest (European Parliament & Council of Ministers, 2002). This perspective differs from that of FUSIONS mentioned above, which includes ready-to-be-harvested crops (but not yet harvested) in its definition of food waste. Regarding the “waste” concept, in the EU legislation, waste is “any substance or object which the holder discards or intends or is required to discard” (European Parliament & Council of Ministers, 2008).

In Romania, the definition of food waste in legal documents differs somewhat from that of the EU, specifically mentioning only food that is degraded and destroyed. The Romanian law defines food waste as “the situation after which food leaves the circuit of human consumption due to degradation and is destroyed, according to the legislation in force" (Law No. 217/2016, Art. 1, point 2) (Romanian Parliament, 2016).

In this book, we used only the term "waste," and not "loss" and "wastage." This choice was driven by the results of the "Team Up for Less Food Waste" project and by the meaning of food waste in the EU (European Parliament & Council of Ministers, 2018) and Romanian law (Romanian Parliament, 2016). Thus, we considered that "food waste" was food intended for human consumption but not consumed by people and which received a different destination (garbage bin or other) along the various stages of the food supply chain (from production to consumption). A similar perspective was adopted in various papers (Bajželj et al., 2020; Talekar et al., 2023; Zhang et al., 2022). Following this logic, fruit and vegetable (FV) waste refers to all FV intended for human consumption that leaves the human consumption circuit, regardless of their ultimate destination.

The purpose of the above presentation of concepts is to highlight the diversity of meanings associated with food waste and the importance of providing a clear definition to avoid misinterpretations. It serves as a straightforward clarification of language for the current book, and we refrain from recommending specific terms, as their suitability may differ according to the context.

In addition to clearly defining food waste, it is important to ensure that similar meanings are present in both the legal framework and in people's minds (this being the second task related to the meaning of the food waste concept, as mentioned at the beginning of this chapter). Misalignments between the legal definition of food waste and stakeholders' understanding of it can jeopardize the implementation of food waste prevention and reduction strategies. Such discrepancies were observed within the "Team Up for Less Food Waste" project, highlighting the need to harmonize the legal and people's definitions of food waste. The one-to-one interviews within the project revealed a different interpretation of food waste compared to the law. A surprising finding was that most participants, regardless of their category, from farmers to consumers and policy makers, considered that FV waste is composed of FV that are thrown away in garbage bins. They did not consider waste those FV that leave the human consumption circuit and receive another destination, such as compost, fertilizer (after being plowed into the ground), and animal feed.

One farmer commented:

I am thinking that [FV waste] is throwing away FV to the garbage ... if we compost them or if we give them to the animals, they are not wasted, not at all .... We do not have any FV waste ... We let them in the orchard, as fertilizer ... We do not throw anything to the garbage bins.

This vision aligns more closely with FUSIONS' definition and the principles of the Food Loss and Waste Standard. The latter is a global standard that provides requirements and guidance for quantifying and reporting on food waste and/or associated inedible parts removed from the food supply chain (Obersteiner & Scherhaufer, 2021). Food waste can occur at each step of the agri-food chain, from production to consumption. However, there are differences regarding the inclusion of specific components or stages in the definition of food waste. While the EU definition of food does not include feed and plants prior to harvest in the "food" category (European Parliament & Council of Ministers, 2002), with their consequent exclusion from the food waste, FUSIONS includes ready-to-be harvested crops (but not harvested yet) in food waste (FUSIONS, 2016). A different perspective (compared to both cited sources) emerged from the "Team Up for Less Food Waste" project. The interviews with farmers showed that their understanding of food and, consequently, food waste, is more comprehensive. Some farmers included immature crops in the food definition (Petrescu-Mag et al., 2024). This difference in perception of what constitutes "food" signals an opportunity for policymakers to adjust the concepts of "food" and "food waste" to align with farmers' perceptions, incorporating immature crops alongside ready-to-harvest ones, thereby addressing food waste more comprehensively (Infographic 2.1).

However, a minority of the interviewed people believed that all food that does not fulfill its initial purpose as human food is wasted. Thus, one transporter said: "*Yes, anything you can no longer sell or derive value from would be considered waste.*" Similarly, when asked if composted FV or those used for animal feed are considered waste, one retailer commented:

> *Yes, everything that is a loss [economic loss] is considered waste.*

Moreover, a mixed opinion was expressed by one processor:

> *Regarding composting, I agree that it constitutes waste. However, when it comes to animals, I partially disagree. In this case, the vegetables will still enter the human consumption cycle through the animal.*

2.2 Food Waste Impact and Statistics

Some of the critical issues in the food sector include food waste, food safety, and environmental degradation. In the past years, the food waste problem has become increasingly important, affecting food security, socio-economic development, and the environment. The increasing population, estimated to exceed 9.1 billion by 2050, will increase food demand (Falcon et al., 2022). Therefore, transitioning to a sustainable food supply chain is necessary. Reducing food waste can reduce greenhouse gas emissions, increase food availability, and improve the use of natural resources (Gardas et al., 2018; Gokarn & Choudhary, 2021; Thorsen et al., 2022). Food production has become increasingly unpredictable and more challenging, with food availability in many areas posing a significant challenge. According to Thyberg & Tonjes (2016), around 11% of the global population is food insecure. Wasted food means wasted nutrients. Globally, on average, wasted food per person per day contains 273 kcal of energy (calories), representing approximately 15% of its global median daily dietary recommended intake (DRI) value, 2.2 mg of iron (24% of the DRI), 323 mg of potassium (12% of the DRI), and 25–50% of DRI of vitamin C, vitamin K, vitamin B6, iron, zinc, copper, manganese, and selenium (Chen et al., 2020). Moreover, natural resources (such as water, energy, or land for food

Infographic 2.1 What is and what is not fruit and vegetable waste

production) are limited and vulnerable to increased socio-economic pressures, and the waste of food also means the waste of these scarce resources that are used to produce food (Piras et al., 2022; Priefer et al., 2016). Food waste is rich in valuable components such as proteins, carbohydrates, minerals, vitamins, and other bioactive compounds, which can be valorized for the extraction of value-added products, such as enzymes, organic acids, and other bioactive components (Ganesh et al., 2022; Sagar et al., 2018). For example, compounds from FV waste can improve the quality and extend the shelf life of meat products (Calderón-Oliver & López-Hernández, 2022), and FV waste can be used as biosorbents for the removal of toxic heavy metals and dyes from wastewater (Patel, 2012).

Some of the impacts of food waste and food production on the environment are:

- *Water pollution and depletion*—the use of pesticides and fertilizers has a negative effect on the surface water and groundwater quality (Craswell, 2021; De Souza et al., 2020). Seventy percent of the world's freshwater is used for agriculture (and 24% of this water is used to produce wasted food/food that is not consumed), being one of the main causes of water resource depletion (Ingrao et al., 2023). According to the FAO (2022), to meet food demand by 2050, water withdrawals will need to increase by 30% compared to current levels. In the EU, 12% of nitrogen pollution in water from agricultural sources is attributed to food waste (Grizzetti et al., 2013).
- *Negative impact on lands*—food waste and production contribute to soil erosion and pollution, nutrient depletion, desertification, and ecological degradation (Thyberg & Tonjes, 2016; Tonini et al., 2018).
- *Greenhouse gas emissions*—food waste (not consumed food) accounts for 8% - 10% of global greenhouse gas emissions [(Mbow et al., 2019) as cited by United Nations Environment Programme (2021)]. In Europe, food waste contributes 4% to the overall European Global Warming Impact [(Scherhaufer et al., 2018) as cited by Obersteiner & Scherhaufer (2021)].

General Statistics:

- 1.13 million tons of edible food products are wasted each day worldwide, representing an average of 178 g of food waste per capita per day (or 65 kg per year). However, differences among countries are notable, ranging from over 500 g/person/day in high-income countries like New Zealand and the United States to 40 g in India (Chen et al., 2020).

– One-third of the food produced for human consumption is wasted globally (Gardas et al., 2018; Heikkilä et al., 2016; Priefer et al., 2016).
– At the household level, the share of food waste in purchased food is also high: 23% of the total weight of food entering the household is wasted in the United States, 17% in Europe, and less than 5% in South Asia and Africa (Chen et al., 2020).
– In Europe, the food waste per capita is between 95 and 115 kg a year (Louis & Lombart, 2018).
– In Europe, in 2020, the sector represented by households produced the highest percent of food waste (53%), followed by processing and manufacturing (20%),

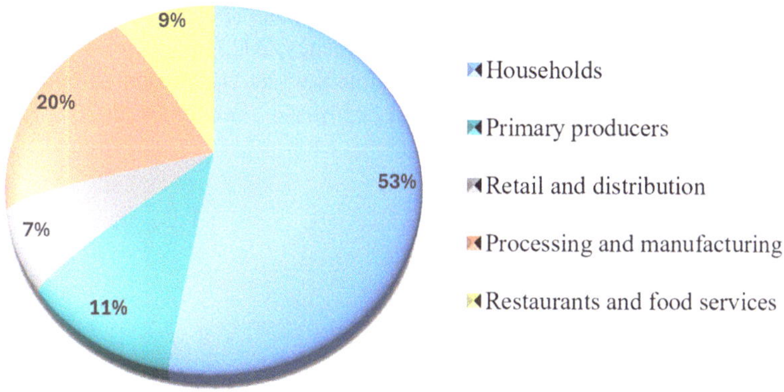

Fig. 2.1 Share of food waste by sector in Europe, 2020

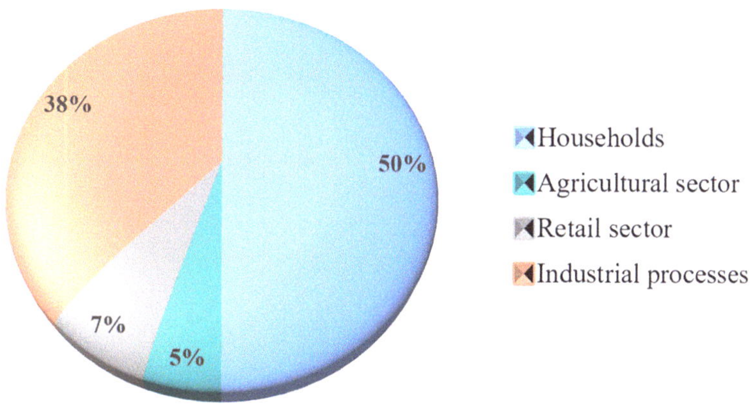

Fig. 2.2 Share of food waste by sector in Romania (European Environment Agency, 2023)

primary producers (11%), restaurants and food services (9%), and retail and distribution (7%) (European Commission. Joint Research Centre, 2023a) (Fig. 2.1).

- Moreover, from the total food waste generated, FV account for the largest share (27% and 20%, respectively). The high percentage is attributed to the fact that fresh products (FV) are perishable, have a short shelf life, and storage conditions vary depending on the type of fruit or vegetable. Therefore, reducing waste from these products is a challenge for the stakeholders along this food value chain (European Commission. Joint Research Centre, 2023a).

Romania Statistics

- Romania generates 5 million tons of food waste every year, ranking 9th in the waste ranking among EU states. Moreover, more than 2.2 billion kilograms of food end up in landfills every year (European Environment Agency, 2023).
- The share of food waste by sector in Romania is presented in Fig. 2.2. The household sector represents the largest share, and according to EEA (2023), this fact

raises some concerns, considering that Romanians spend around 40% of their income on food (European Environment Agency, 2023).

- In Romania, 22% of the food that consumers waste is represented by fruits and 21% by vegetables (*Ghid de consum si pastrare a alimentelor*, 2020).

Chapter 3
Causes of Food Waste and Solutions for Fruit and Vegetable Waste Prevention and Reduction Along the Food Value Chain

Iulia Ajtai, Dacinia Crina Petrescu, Ruxandra Malina Petrescu-Mag, Ioana Cristina Piștea, Carmen Andreea Roba, Oana Adriana Gică, Ancuța Țenter, Valentin Toader, Adina Letiția Negrușa, Daniel Măluțan, and Nicoleta Bican-Brișan

Fruits and vegetables (FV) represent the largest amount of food waste (European Commission. Joint Research Centre, 2023a; FAO, 2015). These products are perishable and have a short shelf life; therefore, reducing waste from this category is more challenging and requires more attention, in some cases, with the loss being inevitable. Implementing efficient strategies should consider all the factors that can influence FV waste and identify specific practices that can improve the management of these products to reduce the waste. Another important issue for this category of food products is related to imperfect-looking FV. The imperfect FV with irregular forms are more likely to be wasted due to the cosmetic standards imposed by distributors/ big retailers and the customers' standards regarding fresh products (Priefer et al., 2016). Customers are less inclined to buy imperfect-looking FV, which results in the waste of these products on retail shelves. Often, retailers have strict cosmetic standards and refuse to purchase imperfect-looking FV, which will lead to more waste in the primary production (farmers) sector. Therefore, FV waste can occur at all stages of the food supply chain for various reasons, and the actions in one sector can influence the FV waste in other sectors.

In this context, the developed strategies must increase coordination along the food value chain. Actions and reduction measures must be implemented in every sector of the food value chain; therefore, it is necessary to have good cooperation among stakeholders to ensure efficient management of food reduction. This involves an exchange of information between sectors, the sharing of best practices, or setting goals and outlining ways to achieve them. Moreover, to improve the efficiency of

I. Ajtai · R. M. Petrescu-Mag · I. C. Piștea · C. A. Roba · A. Țenter · N. Bican-Brișan
Faculty of Environmental Science and Engineering, Babes-Bolyai University, Cluj-Napoca, Romania

D. C. Petrescu (✉) · O. A. Gică · V. Toader · A. L. Negrușa · D. Măluțan
Faculty of Business, Babes-Bolyai University, Cluj-Napoca, Romania
e-mail: crina.petrescu@ubbcluj.ro

I. Ajtai et al. (eds.), *Guidelines for Food Waste Prevention and Reduction: Insights from the Fruit and Vegetable Sector*, https://doi.org/10.1007/978-3-032-14506-2_3

FV's supply chain, communication and collaboration are necessary among partners/stakeholders (Gokarn & Choudhary, 2021).

The first step in identifying the causes of FV waste and developing prevention and reduction solutions was to identify the members of the food supply chain in the case of FV, as well as other stakeholders, collectively considered as members of the food system. First, a list was created based on the review of scientific literature, and members from each category were selected for the project interviews. The list was presented to the interviewees, and they were asked if other stakeholders could be added. The final list included 12 categories: farmers, pickers, processors, transporters, food services (e.g., restaurants and catering), sellers (en-gross and en-detail), storage facilities providers, consumers, waste collectors and processors, other product and service suppliers for FV sector (such as seeds, seedlings, saplings, fertilizers, irrigation systems, plowing works, grafting works, weather forecast services, agricultural IT services), non-governmental organizations (NGOs), and mass media (Fig. 3.1).

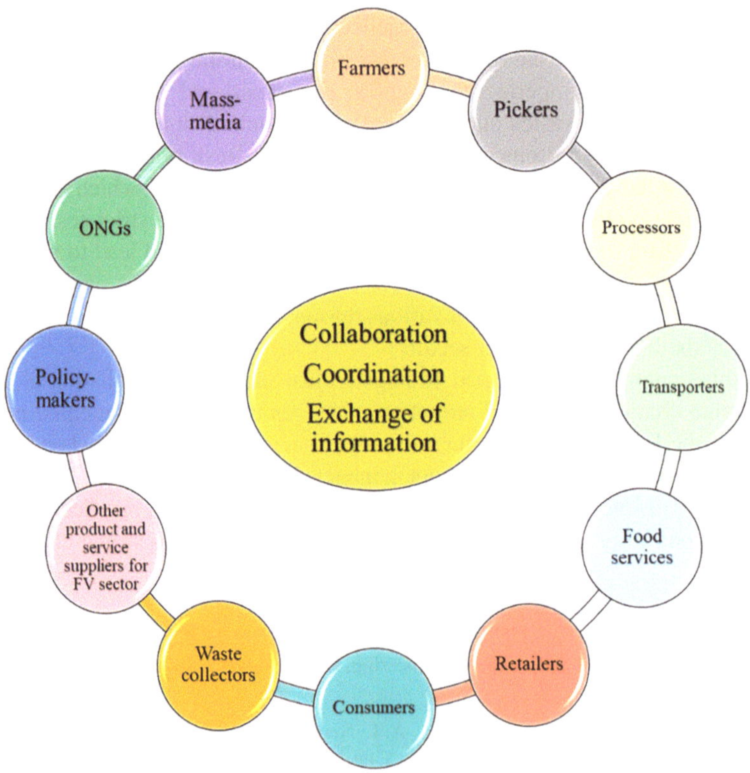

Fig. 3.1 The stakeholders of the food system for FV

3.1 Farmers (Primary Production) and Pickers

Interviews revealed that farmers considered *overproduction* to be one of the main causes of FV waste in the primary production sector (Fig. 3.2). This problem can be generated by low demand for some products on the market compared to production, or by poor planning. The market is unpredictable, with demand for certain products fluctuating from year to year. For instance, the demand for a specific product can be high one year and low in the next. Moreover, favorable weather conditions can increase the amount of FV harvested.

To adapt the production of FV to market demand, farmers must be aware of the types of products customers prefer and estimate the quantities needed. For accurate planning, detailed market studies must be conducted. The types of FV and the quantities that are demanded are not the only essential factors. For instance, knowing the periods in which a certain product will have an oversupply is important in order to avoid overlapping your harvesting with these periods (if possible). It can be useful for local farmers to maintain constant communication with their customers, informing them about the products that will be available throughout the year, so they know what to expect. Another solution can be developing new products (dry fruits, jams, syrup) and processing the oversupply of FV.

Another important cause frequently mentioned by farmers is *aesthetic standards* (Fig. 3.3). The high expectations of customers regarding fresh products will lead to more food waste in the primary production sector. However, consumers are not the only ones who affect the FV waste in this sector. The strict standards imposed by distributors/big retailers (partially because they want to respond to the customer's demand) also contribute to the strict sorting of the products and, therefore, to the waste of the FV that do not meet the standards (Gokarn & Choudhary, 2021; Priefer et al., 2016).

An important factor that can increase the sales of abnormal or imperfect-looking FV is customer education. Farmers can provide more information about the quality and value of their products and participate in various initiatives aimed at raising awareness and educating the population. For instance, ecological FV are more likely to have imperfect forms, but that doesn't mean that their quality is impaired; on the

Fig. 3.2 Solution to reduce FV waste caused by overproduction

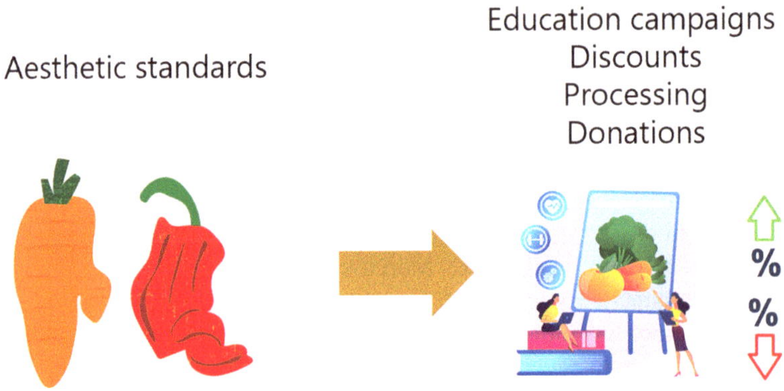

Fig. 3.3 Solution to reduce FV waste caused by aesthetic standards

contrary, these products may taste better and be healthier. According to Louis & Lombart (2018), if the customers are more often exposed to imperfect-looking FV, this will increase the acceptance of these products and purchase intentions. Therefore, more advertisements should include imperfect-looking FV to adjust the consumer's expectations. Of course, other sectors that have a strong connection with the customers and can influence their perception (such as retailers and policy-makers) must get involved in education campaigns or take action—for instance, including more options of imperfect-looking FV in supermarkets and using abnormal FV pictures when advertising. Another solution is to apply discounts for the imperfect-looking FV to donate or process them. Farmers can facilitate the donation process by creating collaborative platforms or establishing relationships with organizations that facilitate donations. Despite the attractiveness of processing, the investment costs and difficulties in finding a market for the processed products prevented some farmers from processing their FV surplus. Others could invest in processing facilities, produce jams, powder, or other products, and increase the profitability of their business. One farmer talked about her experience:

> *We used to throw away large quantities of our fruits because of the lack of time to process them or low demand compared to the offer during the peak season. I was unhappy with this situation, so I started preparing jams and other preserves and sold them. I calculated, and I concluded that the earnings cover the costs, repay for the time spent, bring me profit, and save my work and other resources from being thrown away into the trash.*

The *low market price* was mentioned by farmers as another cause of FV waste (Fig. 3.4). In some cases, the market price of the products will not cover the cost of production, making the harvesting uneconomical and unjustified. This problem can be attributed to the fact that imported products abroad are cheaper or to the fact that large farmers sell their products at lower prices. However, small local farmers cannot afford to do so because the low prices do not cover the cost of harvesting, so they prefer to leave the FV unharvested in the field. The case of organic FV is even more problematic because the production cost is usually higher, more work is needed, losses are higher, and therefore the final price for these products is higher. The lack

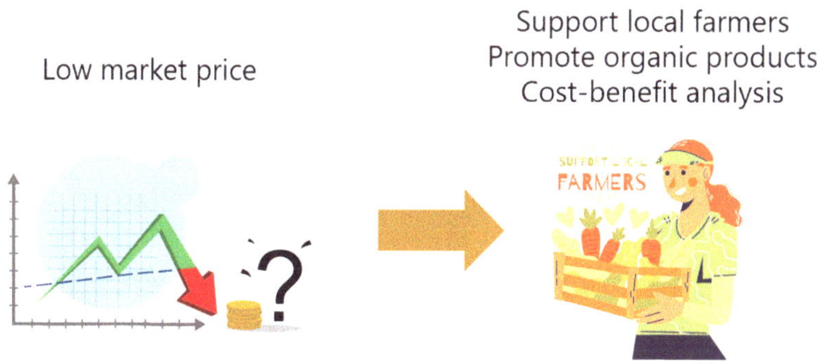

Fig. 3.4 Solution to reduce FV waste caused by low market price

of education and knowledge regarding the properties of ecological products, as well as the associated efforts and costs to produce them, often leads customers to opt for cheaper, conventional products.

A solution to the low-price problem is to encourage people to support local farmers and to advertise local products. It is important to remind people that supporting local farmers can improve the local economy, keeping money within the community. Promoting organic products can also influence customers' purchasing decisions. Informing people about the benefits of organic FV and the differences between conventional and organic products can increase the purchases of local organic products. Moreover, farmers should conduct a detailed analysis of production costs and market prices for the FV they want to grow and perform a cost-benefit analysis to determine the profitability of their business, as well as other relevant analyses. This information will allow them to adjust their technologies and market strategies. Another solution is for the government to improve the incentive schemes that support food production.

Farmers often acknowledged during the interviews that the *long supply chain* (Fig. 3.5) increases the time it takes for the products to reach the consumers from the moment they are harvested. Given that the FV are perishable products, they must reach the consumer in the shortest time possible. The intermediaries between producers and consumers increase the risk of spoilage due to potential transportation delays, poor storage conditions, poor communication, and a lack of coordination.

Furthermore, the supply chain is far from perfect, and the farmers encounter many problems that must be overcome. For instance, the requests from the big retailers are hard to fulfill, according to one of the interviewed farmers:

> *Stop asking for tomatoes in December, because it's not the tomatoes season…. the reasons why we waste FV are: the biggest, because I don't have enough marketing channels; because I, a small producer, go to the supermarket, no matter who it is, and tell them: « I produce the better, cleaner, tastier tomato», but he doesn't care. He is interested in two things: the price, which should be at most the stock market price, not higher; and secondly, constancy. That is: « if you can give me 300 kilos of tomatoes every month from January to January; if you can't give me that, and we're only discussing seasonal situations, I'm not interested in working with you. ».*

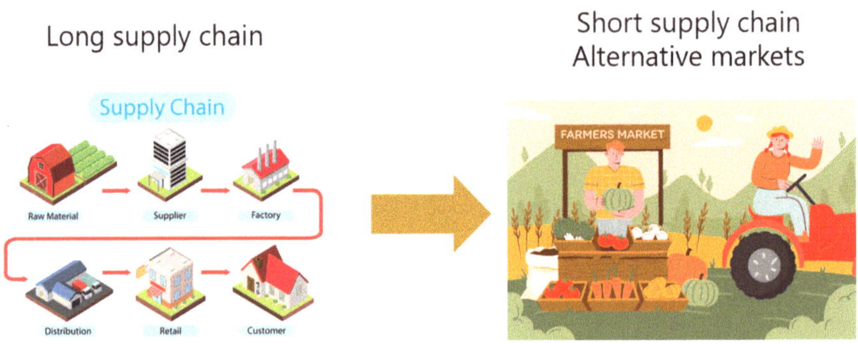

Fig. 3.5 Solution to reduce FV waste caused by the long supply chain

Therefore, farmers face difficulties establishing constant collaboration due to seasonal constraints.

Farmers also mentioned that there are not enough marketing channels, particularly local marketing channels or farmers'markets, that would improve the trade between producers and consumers.

In this context, the best solution is to reduce the supply chain length and use a short supply chain to have a direct connection between farmers and consumers. By having a direct marketing of products, the transport distances between producers and consumers are reduced, and, in this way, the risk of spoilage is reduced (Priefer et al., 2016), as participants also acknowledged.

One farmer explained what he did on his farm:

> *The best solution was to reduce the supply chain, so that, at most, one day from harvesting to the moment the product reaches the consumer. Then, within the farm, we reduced the time between harvesting and the moment when the products reach the cold storage. I think we lack a platform where we can make our products available, like an online market somehow, where someone comes and buys them, because in this way, we could reduce losses.*

Moreover, diversifying the sales channels by identifying alternative markets could be a good way to establish new collaborations. For instance, the farmers can sell the FV directly to restaurants and independent shops, can search for online platforms (marketplaces) to sell their products, or establish collaborations with other farmers to create Farmers'Market Cooperatives (Milford et al., 2021).

Farmers also explained that *improper technologies, equipment, and storage conditions* (Fig. 3.6) can lead to increased waste of FV. Sometimes, mechanized systems can affect and damage the FV during harvesting. This can be caused by faulty technologies and methods or a lack of workers' knowledge regarding harvesting methods. However, due to labor shortages and high labor costs, it is common for farmers to adopt mechanized systems that can ultimately reduce costs (Hamilton et al., 2022). Moreover, some FV have a shorter shelf life than others. Therefore, farmers must pay special attention to the most perishable products and provide them with proper storage conditions. The construction of proper cold storage facilities

Fig. 3.6 Solution to reduce FV waste caused by improper technologies, equipment, and storage conditions

involves high costs, but they will reduce the cost caused by FV waste. The storage of FV should be done in cold storage facilities or using cooling chambers. The use of wooden or fiberboard boxes can also reduce FV spoilage.

Other solutions suggested during the interviews are keeping the equipment well-maintained and using efficient technologies with high profitability. Staying informed and being up to date about the best harvesting technologies and techniques can be a way to reduce damaged crops (STREFOWA, 2019d), and interviewed farmers mentioned this fact. According to the interviewed people, training warehouse staff on product handling and knowledge transfer among farmers also proved to be an efficient solution (Infographic 3.1).

3.2 Processors

The FV waste in the processing sector can have several causes, which will be further discussed. The main cause mentioned by the processors interviewed in the project is the overstocking of products used for processing (Fig. 3.7). Not knowing or being unable to accurately estimate the capacity of processing installations and technologies can lead to a surplus of FV. Additionally, a labor shortage can be a problem when large quantities of products need to be handled, which can increase processing time.

One processor explained:

> *In case we don't think about the fact that we can process a certain quantity, and we have purchased more...we don't have enough space, or our processing line can't process as much as we purchased. That would be one cause of FV waste.*

Therefore, because of their perishability, some of the FV will spoil before entering the processing system. This can be prevented by a thorough assessment and

Infographic 3.1 Causes and solutions for FV waste in the primary production sector

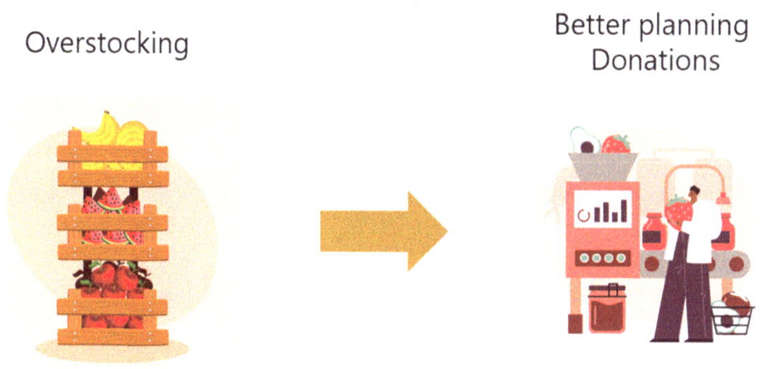

Fig. 3.7 Solution to reduce FV waste caused by overstocking

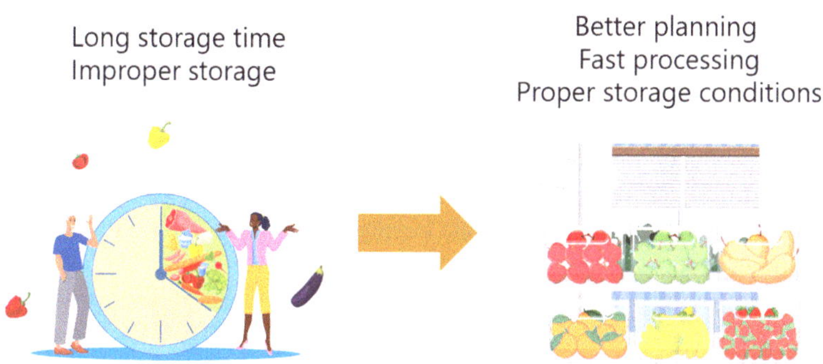

Fig. 3.8 Solution to reduce FV waste caused by long storage time and improper storage

understanding of the processing system used, as well as the amount of FV that can be processed within a certain timeframe. The stocks should be made depending on the type of FV and their perishability. For example, it is possible to stockpile products with a long shelf life, but it is not advisable to purchase large quantities of highly perishable food items. Donations of excess FV can also be made to food banks or various associations.

Moreover, the *long storage time and improper storage* can lead to FV waste (Fig. 3.8). Processors repeatedly mentioned situations where the quantities of FV exceed the production capacity or the presence of technical problems with the production line, which can delay FV processing. These problems will increase the storage time, leading to the spoilage of the more perishable FV. Additionally, storage conditions should be suitable in terms of temperature and humidity, particularly when large quantities of FV are present.

To avoid these problems, the processing of FV should be done right after purchasing or as soon as possible. Another solution is to avoid overstocks by better planning. The storage should not be overloaded with FV, and it should not surpass its recommended capacity. By implementing cold storage facilities or other

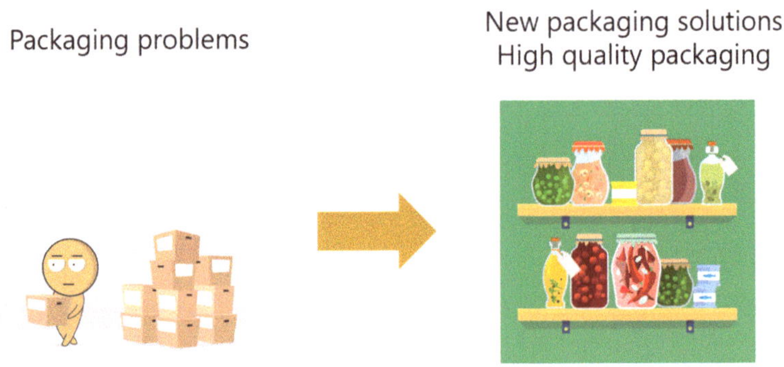

Fig. 3.9 Solution to reduce FV waste caused by packaging problems

technological innovations, such as temperature monitoring, the shelf life of FV can be increased.

Another cause of FV spoilage is the *packaging problems* (Fig. 3.9). Some of the processors complained about the faulty packaging, which resulted in the loss of processed products. For instance, items sealed in jars with faulty lids resulted in product spoilage. Therefore, it is important to implement new packaging solutions, thoroughly check the quality of the packaging that is used, or invest in high-quality packaging in order to avoid such problems.

The *poor quality of the products* delivered by suppliers is another problem in the processing sector mentioned during the interviews (Fig. 3.10). If the FV are considered to not be suitable for processing, they are sorted with the purpose of eliminating the damaged, spoiled, or irregular FV (IRCEM, 2021). Moreover, the FV are also subjected to strict sorting due to the standardization regarding the size and quality of products. According to (Cane & Parra, 2020), the quality standards that must be implemented cause a lot of waste in the processing sector. Therefore, processors should be cautious when selecting suppliers to collaborate with and regularly inspect the quality of the products delivered. Another solution to reduce FV waste, suggested by processors, is to use the products that cannot enter the processing line for other purposes. In this way, new products can be obtained, diversifying the product options and creating new business opportunities (Gardas et al., 2018). For instance, the canned FV need to respect certain standards, and, therefore, the ones that are eliminated from this production line can be used to produce other products such as sauces, jams, syrups, or fillings. The processing companies can also develop collaborations with food banks to donate FV for human consumption, or they can donate to different system players that need feed for animals.

Poor processing infrastructure or *technical failures* can also cause production line waste, according to several processors (Fig. 3.11). The main factors that will influence the FV waste are related to the inconsistency of manufacturing processes, outdated and unsuitable methods and techniques of processing (FAO, 2019; Gardas et al., 2018). This waste can be reduced by improving processing technologies, optimizing processing methods, or utilizing efficient equipment. However, one

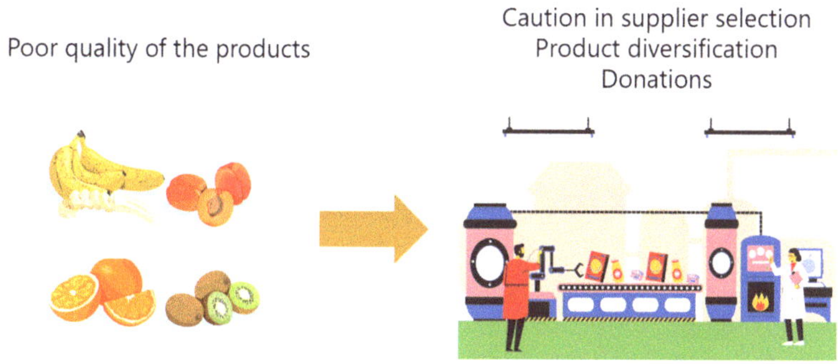

Fig. 3.10 Solution to reduce FV waste caused by the poor quality of the products

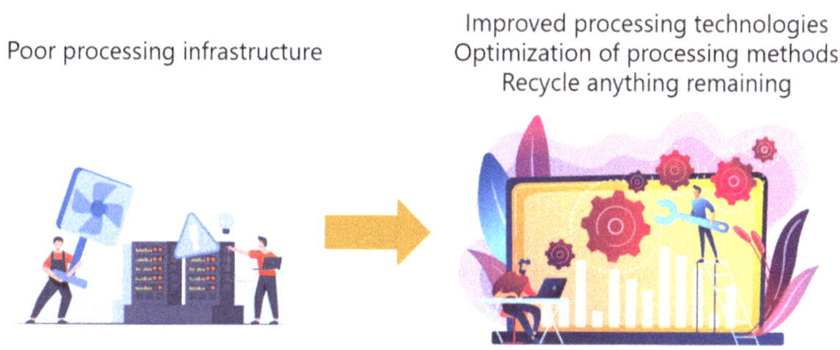

Fig. 3.11 Solution to reduce FV waste caused by a poor processing infrastructure

processor mentioned that investing in better processing technology implies more costs for them, and sometimes these costs exceed the ones related to the FV waste along the production line. Although in the long term, the waste reduction is to the benefit of processors, they often choose not to implement expensive measures. In this case, another solution to prevent FV from ending up in the landfill is to recycle any remaining materials. The FV that exits the production line as waste can be used as livestock feed, fertilizer, or compost.

One processor mentioned the *lack of knowledge* regarding the best methods and techniques as another cause for FV waste in this sector (Fig. 3.12). It is important to understand all the processes that occur along the production line in order to implement the most effective strategies. By staying up to date with technological advancements and being informed, the processors can find the best solutions to maximize product utilization and reduce waste. Moreover, employees must be regularly trained and provided with information on best practices (Infographic 3.2).

A processor said:

> *I remembered that in the first year of processing, we had products that fermented because we did not master the pasteurization technique, but in the meantime, we got more precise pasteurization equipment, and we also understood the process better, and we no longer had such problems.*

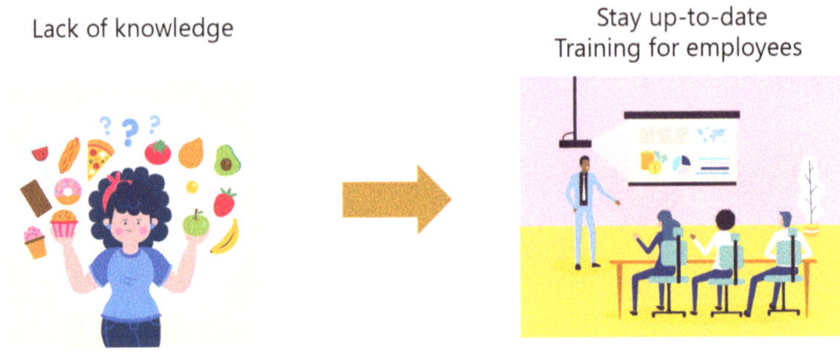

Fig. 3.12 Solution to reduce FV waste caused by a lack of knowledge

3.3 Transportation

The interviewed transporters considered that the main cause of FV waste in this sector is the *improper transportation conditions* (Fig. 3.13). These conditions primarily refer to the temperature and humidity required to maintain the freshness of the FV. When FV are transported in non-refrigerated trucks during the summer, the spoilage rate will be higher due to their perishability and the high temperatures. Many of the interviewed transporters acknowledged that this is a significant problem or an issue that can be addressed by implementing adequate transport conditions, such as refrigerated transport services or fresh air vents. These systems will also prevent temperature fluctuations that can cause the spoilage of FV. They are also useful for maintaining a constant temperature during winter, when the cold can cause the FV to freeze during transportation. Another aspect that must be considered is the fact that different FV need different temperature conditions. However, these technologies imply high costs, as not all transporters can afford to implement them.

Another important cause highlighted during the interviews was the *improper packaging* during transportation (Fig. 3.14). The packaging must be selected according to the FV that are transported. Some FV are not so sensitive (such as apples or potatoes) and don't need extra care when transported. However, the more sensitive ones (such as peaches or tomatoes) require special packaging to avoid damage. It is essential to select packaging materials that prevent FV spoilage, avoid overloading the vehicle with FV, and package them in a way that prevents them from bumping into each other. Small boxes or crates should be used for more sensitive products instead of sacks to avoid pressure on the FV.

Transporter:

> *...If they are not properly packed in boxes. Then indeed, during delivery, they can bump into each other and get damaged. Yes, for example, we actually experienced this with our tomatoes, which are very sensitive. We put them in plastic boxes, and they took the shape of the box; as a result, we could no longer sell them. Yes, they have a very thin shell, and they are affected immediately, and that is why they must be put in cardboard boxes. Packaging, yes, is a problem during delivery if they are not properly packed. It is possible to bump into each other, to be damaged.*

Infographic 3.2 Causes and solutions for FV waste in the processing sector

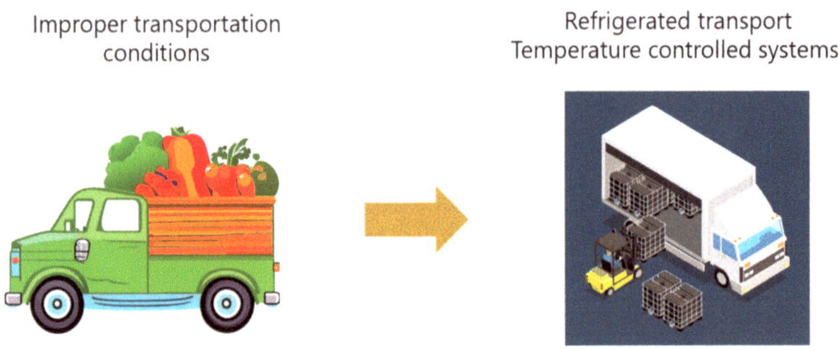

Fig. 3.13 Solution to reduce FV waste caused by improper transportation conditions

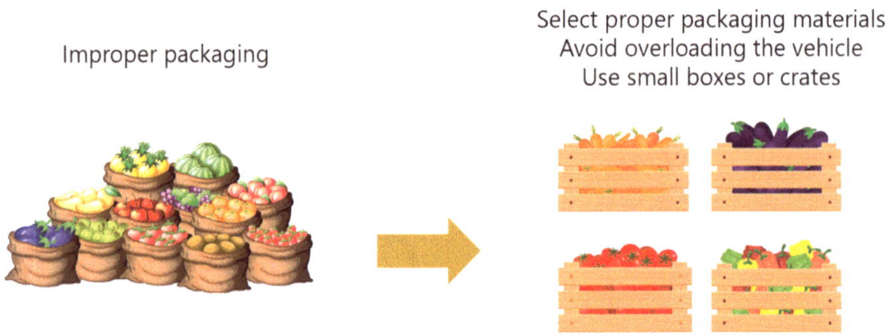

Fig. 3.14 Solution to reduce FV waste caused by improper packaging

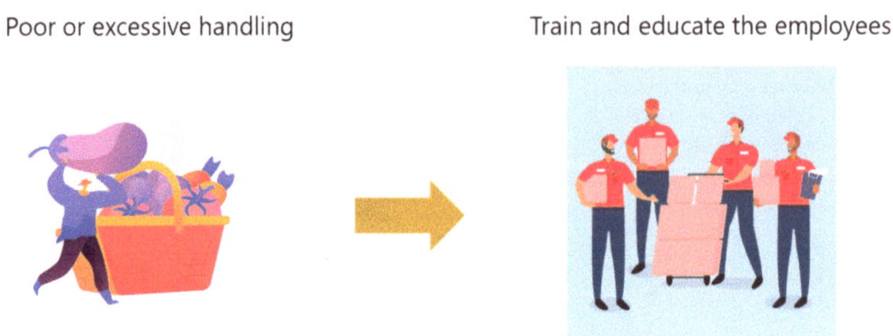

Fig. 3.15 Solution to reduce FV waste caused by poor or excessive handling

Transporters also showed that *poor or excessive handling* can damage the FV (Fig. 3.15). The mechanical damage to FV that can occur during transportation affects the visual quality of these products and reduces their economic value. This can cause the retailers or the customers to refuse the products upon receipt. Rough handling by market retailers or other workers can cause physical damage to the fresh products and accelerate the spoilage of these products (Al-Dairi et al., 2022). Therefore, employees should receive education on these aspects and undergo

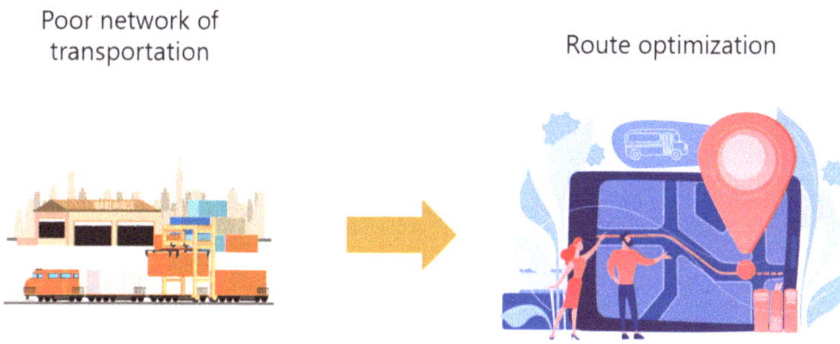

Fig. 3.16 Solution to reduce FV waste caused by a poor transportation network

comprehensive training to acquire knowledge on the best methods for accomplishing various tasks during FV transportation and delivery. In this way, they will acquire the necessary skills for handling FV, following protocols, and quality control measures.

Another aspect mentioned in the interviews was the *poor transportation network*, which led to FV waste (Fig. 3.16). Cheaper transportation channels are often used for economic efficiency, which can cause products to travel longer distances, thereby increasing the risk of spoilage. The poor transportation infrastructure, such as rough roads, increases mechanical vibration and shock during transport, which has a negative impact on the quality of fresh products. Therefore, time and distance must be considered when planning the transport routes (Gardas et al., 2018; Mena et al., 2011). The best solution to avoid waste due to a poor transportation network is the optimization of transport routes, including quick delivery and short distances from factories or farms to supermarkets or customers. This will reduce the time FV spend in transit, reducing the exposure to the factors that cause quality deterioration. Moreover, this solution also reduces transportation costs, providing an economic advantage for producers.

Similar to other members of the FV supply chain, processors observed that *poor coordination* between the involved parties, such as farmers, transporters, and customers, can cause the FV to spoil before delivery (Fig. 3.17). If farmers and producers do not consider transportation logistics in the planning process, they risk not being able to deliver all their products to the market before they spoil. Moreover, inefficient communication between the parties involved can create delays in delivery. End-users have high expectations, demanding high-quality products, prompt delivery, and convenient communication channels. If the time of delivery is not respected, the FV will spend more time in transit, causing a deterioration in quality, which can also prompt the recipient to reject the shipment (*Ghid de consum si pastrare a alimentelor*, 2020). Therefore, poor logistical coordination is an important factor that must be addressed and improved. In this case, the FV waste can be reduced by establishing effective communication channels, proper coordination and planning, offering discounts, or donating the FV that are rejected by recipients (Infographic 3.3).

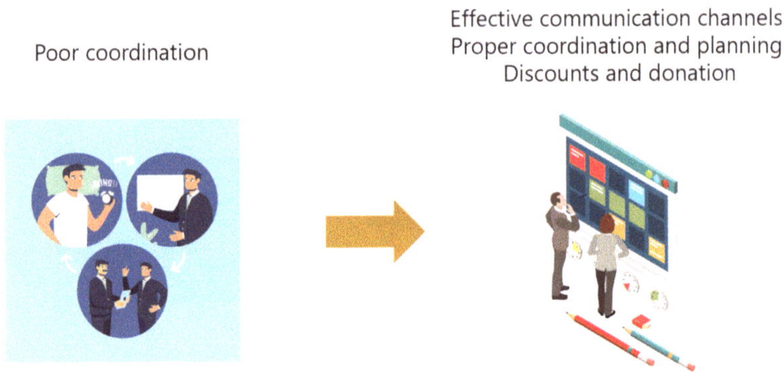

Fig. 3.17 Solution to reduce FV waste caused by poor coordination

Transporter:

We have several options. ...For example, we have a box of vegetables, which had to be delivered to a certain person. That person forgot to announce that they were out of town or whatever. We have some customers to whom we sell a box of vegetables at a reduced price. This is the first option. Or the option of offering, for example, to the Center for Palliative Diseases. We offer the vegetables to third parties, as well as to certain families in need and disadvantaged families that we are aware of. At the end, if we still have some FV left and they are spoiled, they are used as compost.

3.4 Food Services

Interviews revealed that poor planning or a lack of planning in the food services sector is one of the main factors influencing FV waste here (Fig. 3.18). In this case, planning encompasses various aspects. One of them is planning the quantities of products that need to be purchased to prepare the food on the menu for a given period. Of course, this depends on the demand. The number of customers varies, making it challenging to estimate the quantity of food that will be prepared daily. This can lead to overstocking of the raw products, which require sufficient space for storage and proper storage conditions. Therefore, effective stock management must be implemented, improving sales estimations based on past experience, buying raw products more frequently and in smaller quantities, or ensuring sufficient storage space with optimal conditions. Another solution to avoid overstocks (whether we are talking about raw products or prepared food) is to apply dynamic pricing, including the introduction of happy hours at the end of the day. This approach involves adapting pricing structures in response to real-time demand and inventory levels. The businesses can offer discounts or promotions on food items that are approaching the end of their shelf life.

Another aspect is planning the kitchen operations (Heikkilä et al., 2016). The order in which the products are processed is important in the FV waste generation,

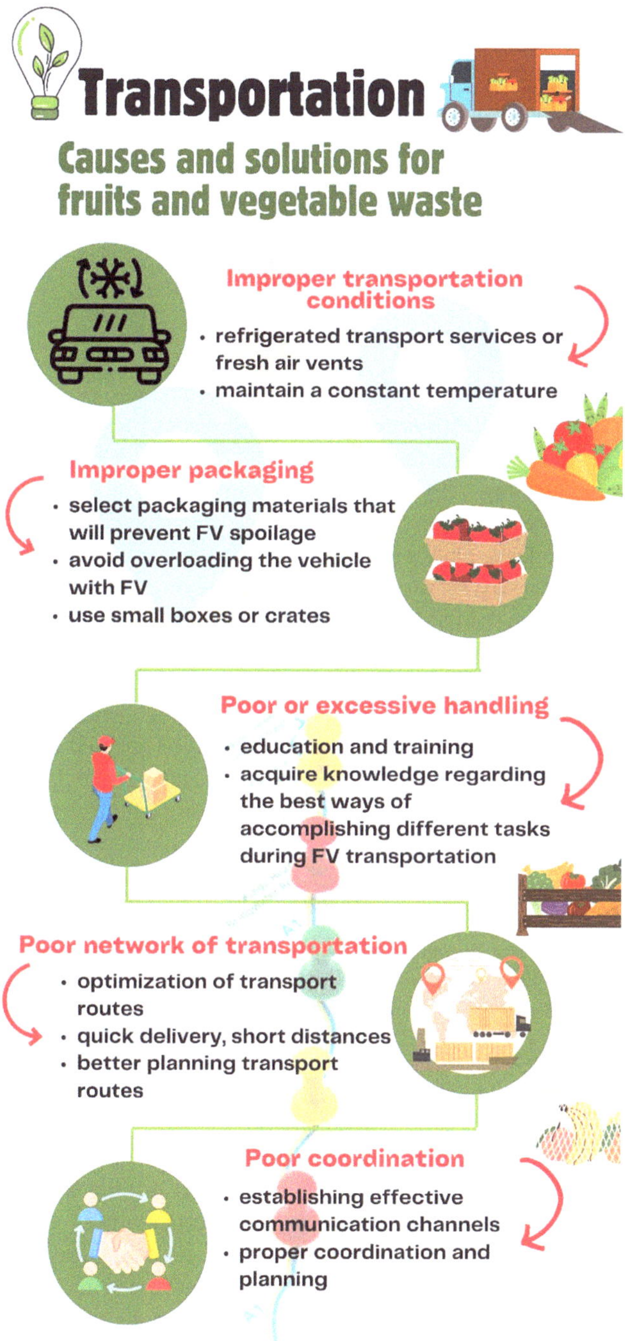

Infographic 3.3 Causes and solutions for FV waste in the transport sector

Fig. 3.18 Solution to reduce FV waste caused by poor planning or lack of planning

especially when dealing with overstocks. The first products that must be used for food preparation are the oldest ones, which have been stored the longest and have a higher risk of spoilage. Only after these products are used can the fresher ones, purchased more recently, be used for cooking. Applying the "first-in-first-out" rule can prove effective in reducing FV spoilage before consumption. Moreover, establishing a routine in the cooking process for different dishes will help employees avoid mistakes, which can result in food waste.

The planning process includes the menu as well. The menu should be designed in a way that allows flexibility of the dishes, should be adapted to the season, and should include dishes that can be cooked efficiently without generating high amounts of food waste (Vizzoto et al., 2021). For instance, fruits that are about to spoil or lack the desired quality for a certain dish can be used to make smoothies or fresh drinks.

A chef suggested:

> *In the restaurant where I work, I think that if more freedom were allowed in terms of the menu, certain products that are about to expire or lose their validity could be better used. If the focus was on creativity and using them that day, even if you didn't have something on the menu, they would not end up in the trash. Yes, I think that would be a solution from my point of view: to use seasonal fruits and vegetables and be willing to find an original and quick use for them on the day, so that they don't end up in the trash. Yes, but that is sadly missing.*

Interviewed chefs and restaurant managers mentioned that the *lack of knowledge* also generates kitchen waste (Fig. 3.19). Employees lacking experience or those disinterested in the food waste problem may inadvertently use inefficient practices or make cooking mistakes. These problems can be addressed through regular training to equip employees with the necessary knowledge and skills, raising awareness about the consequences of food waste, and educating them on various aspects. These aspects should include how to avoid over-trimming fruits and vegetables, strategies to maximize the utilization of each product, and the art of utilizing all parts of the ingredients. Additionally, training should extend to the efficient

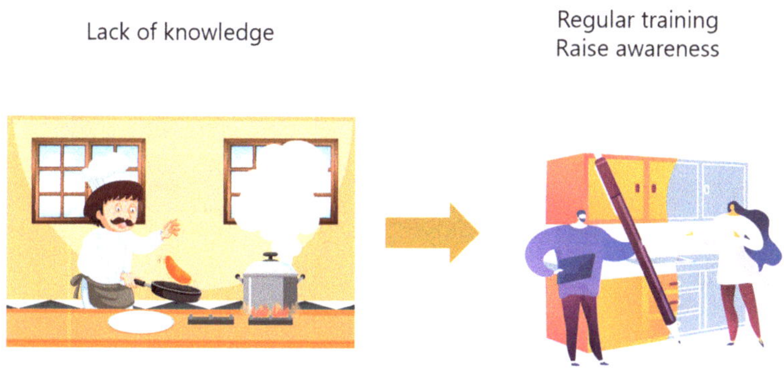

Fig. 3.19 Solution to reduce FV waste caused by a lack of knowledge

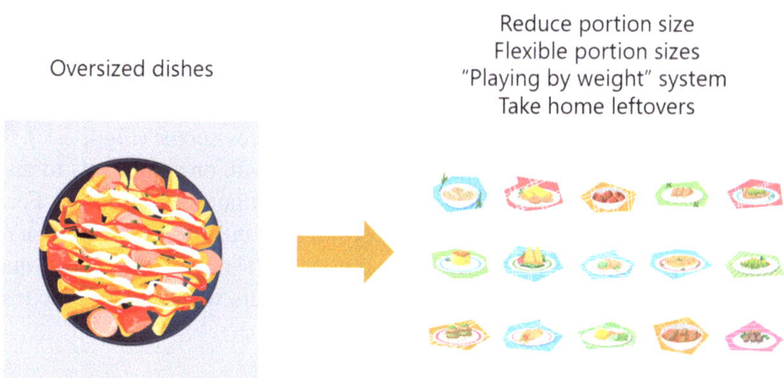

Fig. 3.20 Solution to reduce FV waste caused by oversized dishes

preparation of leftovers, emphasizing creative and sustainable cooking methods to minimize waste (Vizzoto et al., 2021). The chefs can think of innovative purposes or uses for products that may have lost their pristine quality but remain suitable for consumption.

Interviewees noted that oversized dishes are another issue that contributes to plate waste (Fig. 3.20). One solution could be to reduce portion sizes; however, this approach might not resonate universally with customers, as not everyone eats the same amount of food at a meal. Thus, flexible portion sizes could be a better option. Providing options for flexible portion sizes enables customers to tailor their orders to their individual hunger levels and preferences. This approach not only minimizes plate waste but also enhances customer satisfaction by catering to a diverse range of dietary needs. The buffet restaurants also have the potential to generate large amounts of food waste, as customers tend to take more food than they can eat, often overestimating their consumption capacity. According to Priefer et al. (2016), this can be prevented by placing visible signs to remind the customers to take only as much food as they can eat or by implementing a "playing by weight" system, where

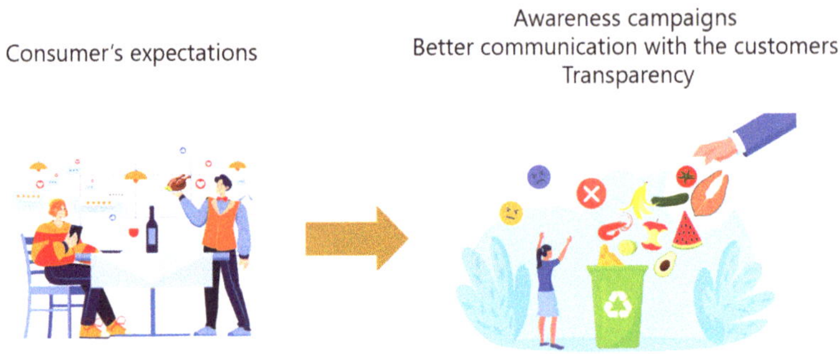

Fig. 3.21 Solution to reduce FV waste caused by consumers' expectations

customers are charged based on the weight of the food they select. Businesses should also actively encourage customers to take home leftovers, providing take-away containers or placing them in accessible areas within the restaurant.

Interviewees draw attention to the fact that consumers' expectations for top-quality products, in terms of both taste and appearance, as well as a wide variety of options, can also lead to food waste in the food services sector (Fig. 3.21.). They expect fresh products of the best quality, placing pressure on restaurants to discard items that do not meet stringent standards or exhibit minor imperfections. For certain restaurants that take pride in the excellence of their cuisine, maintaining a reputation for delivering top-notch food is paramount. Therefore, stringent quality control measures become a necessity to retain customers who associate the restaurant with consistently superior food offerings.

One of the interviewed chefs explained:

> *We value the freshness and quality of our dishes, so each ingredient must be as fresh as possible, as good as possible from a qualitative point of view. Therefore, we cannot afford not to waste anything. Unfortunately.*

Moreover, the discard practice is more noticeable in buffet restaurants where customers anticipate a broad selection, even during the later hours of the day, leading to unnecessary food waste (Heikkilä et al., 2016).

Awareness campaigns should be implemented to inform customers about the environmental impact of food waste and encourage more mindful consumption. Better communication with customers may also be beneficial. The business must be transparent about its commitment to sustainability and quality, explaining its efforts to minimize food waste and how minor imperfections do not compromise the overall taste or safety of the dishes.

The poor quality of products delivered by suppliers is another issue mentioned in the interviews conducted during this project (Fig. 3.22). Restaurants typically do not use all the delivered products on the same day; instead, they will store them for a period of time. Thus, if the products are not fresh or exhibit signs of withering, they will spoil before use. The raw product quality will also influence the food quality, which is an important factor in meeting customers' expectations. If the food has

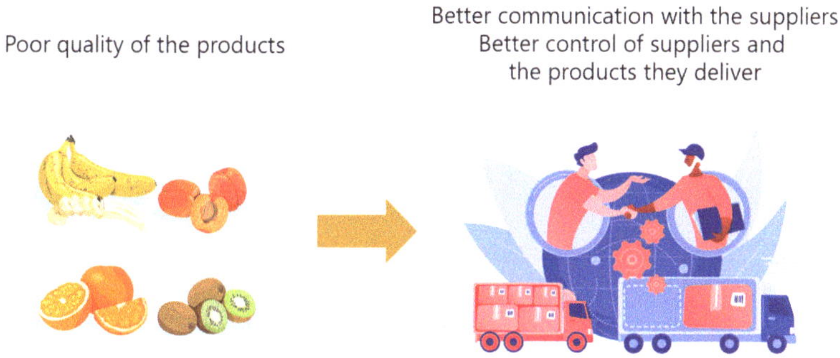

Fig. 3.22 Solution to reduce FV waste caused by the poor quality of the products

a low quality (regarding the taste or appearance), the customers will not eat the dish, and the leftovers end up in the trash. Some of the solutions involve better communication with suppliers or improved control over the products they deliver. Establishing enhanced communication channels with suppliers ensures a clear understanding of the expectations regarding product quality, delivery timelines, and freshness standards. Supplier control may include regular quality checks, audits, and ongoing collaboration to promptly address any issues that arise (Infographic 3.4).

3.5 Retailers

The lack of cold storage is one of the main causes of FV waste, as mentioned by the interviewed retailers (Fig. 3.23). Cold storage plays a crucial role in preserving the freshness and quality of perishable products. The FV are also sensitive to temperature fluctuations. Without cold storage facilities, the products' shelf life is drastically reduced, accelerating the spoilage process. Despite some small retailers acknowledging during interviews that having cold storage could reduce food waste, the associated expenses pose a considerable barrier. Many of these businesses find it financially challenging to make the necessary investment in cold storage infrastructure. A retailer pointed out:

> *Waste can happen due to temperature conditions that are not ideal for fruits and vegetables… Many fruits and vegetables, not just in our case, but throughout the retail space, are placed in an area alongside other products, not just fruits and vegetables. Therefore, the temperature is not ideal. There should be a much lower temperature than in the rest of the sales area. However, due to the actual space conditions that you find in the store, there is no way to maintain a certain temperature of 16 degrees in the fruit and vegetable area and 20 degrees everywhere else in the store - it's physically impossible, or the cost for something like that would be exorbitant.*

To eliminate this problem, the solution involves implementing technologies that extend product shelf life, including cold storage, temperature monitoring, ensuring adequate air circulation, and intelligent packaging, all aimed at improving product

Infographic 3.4 Causes and solutions for FV waste in the food services sector

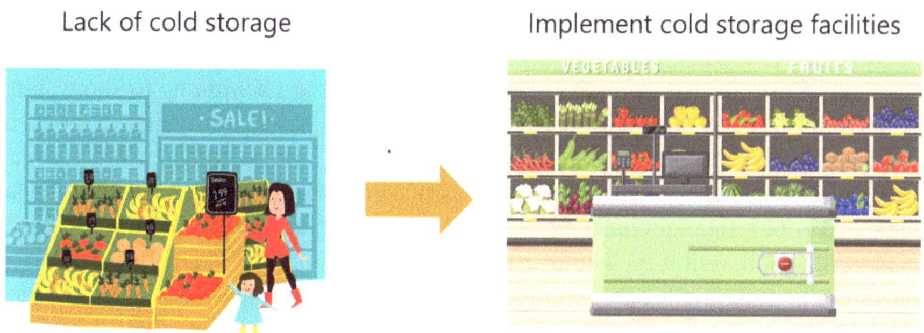

Fig. 3.23 Solution to reduce FV waste caused by a lack of cold storage

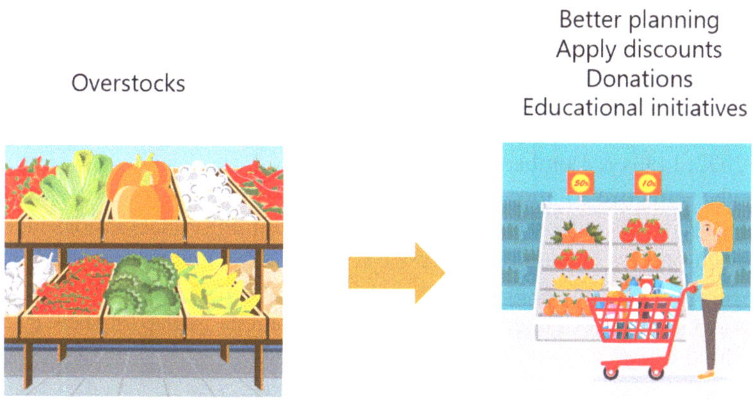

Fig. 3.24 Solution to reduce FV waste caused by overstocks

quality and freshness. The cold storage maintains controlled low temperatures and regulates humidity levels, enhancing the overall quality and freshness of FV (STREFOWA, 2019c).

Another significant factor contributing to FV waste in the retail sector is related to overstocking (Fig. 3.24). The retailers mentioned that this problem can have several underlying causes. One is the inadequate estimation of the quantities of FV that the store manager needs. Seasonal variations and the consumers' preferences for certain products can create fluctuations in market demand that are difficult to predict. Another one is pricing dynamics. High prices on certain items may deter customers, leading to lower sales and, consequently, overstocking situations. Moreover, customers demand a large variety of products to be available all the time; therefore, retailers overstock their stores to satisfy this demand.

To avoid overstocking, managers should conduct market studies to forecast demand for different products. Identifying consumer preferences and market trends can improve the accuracy of product demand predictions. Another strategy could be to buy smaller quantities of products more frequently, rather than purchasing large quantities in bulk. This way, orders regarding the quantities of FV that are purchased can be adjusted more easily according to sales, avoiding overstocking. However, there are instances where overstock cannot be avoided. In this situation, a practical solution is to apply discounts for the FV that are beginning to show signs of declining quality. Implementing dynamic pricing based on available stock is another solution. For instance, retailers can create specific sections known for featuring discounted items within their stores or supermarkets. When creating these dedicated zones, retailers must ensure that they are properly placed and well advertised, with good visibility within the store. The discounts create an incentive for customers to make swift purchases, but also accelerate the sale of these items before they get spoiled and contribute to unnecessary waste. Discounts could also be extended for abnormally shaped or imperfect FV to draw customers' attention to these types of products. Moreover, introducing abnormal-shaped or imperfect FV more often in the store's offerings and advertising them can be a valuable initiative. Identifying the most effective marketing channels, such as in-store signage, social media campaigns, or targeted promotions, plays a crucial role in successfully integrating these products into the mainstream consumer consciousness. Such tactics expose consumers to these types of products and increase their acceptability (Eičaitė et al., 2022; Louis & Lombart, 2018).

Another effective solution to avoid FV waste suggested by retailers is to donate the excess products to food banks or local NGOs. Building collaborations facilitates an efficient process for donating excess FV. Regular communication and coordination with these organizations ensure that surplus produce reaches those in need promptly. One retailer talked about their collaborations as part of their efforts to reduce food waste:

> *We work with food banks…And we call this food bank daily, which comes and takes all the fruits and vegetables that can no longer be sold under normal conditions by us. This is our decision. We do this sorting; they obviously can be consumed very well. And this food bank takes these fruits and vegetables and sends them to the various beneficiaries for free.*

Some retailers organize educational initiatives to raise awareness about food waste and provide consumers with valuable information on the shelf life of FV. Clear messages about the extended usability of FV, even if they exhibit slight imperfections or are approaching their expiration date, can significantly impact consumer behavior.

Again, *the poor quality of products* delivered by suppliers was indicated to cause FV waste (Fig. 3.25). The FV possess a limited shelf life that varies based on their specific nature. Generally, fresh products have a limited window of time during which they can be purchased before their quality deteriorates. Once these products reach the store, there is a considerable amount of time during which they remain on the shelves before being bought by consumers. If these products are already near the end of their shelf life upon arrival at the store, there is a heightened risk of spoilage, thereby contributing to wastage within the retail sector. Additionally, if the products are of poor quality (or their aesthetic characteristics do not align with consumers' expectations), customers are less inclined to purchase them even if they are good for consumption. This results in a prolonged shelf life for these products in the store, increasing the likelihood of spoilage and contributing further to overall FV waste.

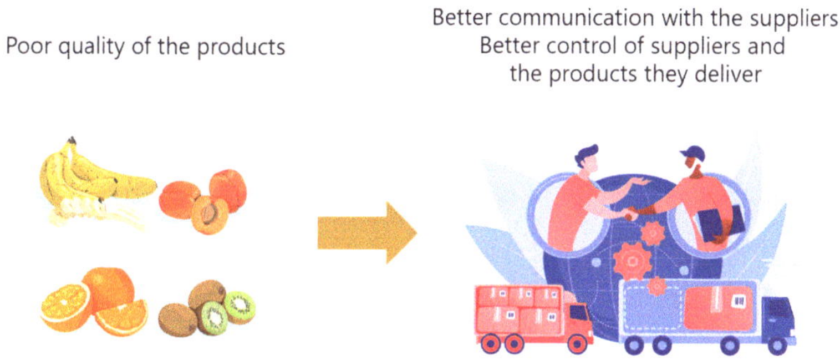

Fig. 3.25 Solution to reduce FV waste caused by the poor quality of the products

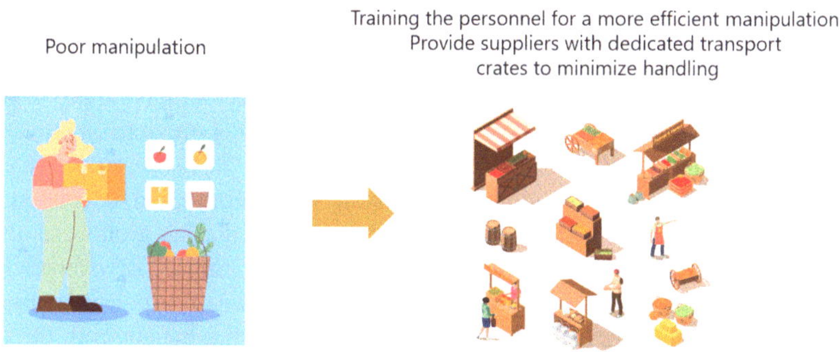

Fig. 3.26 Solution to reduce FV waste caused by poor manipulation

Addressing the timeliness and quality of product deliveries is highly important to minimize waste and enhance overall efficiency in the retail supply chain. The retailers should conduct quality control checks upon the arrival of products at the store. This includes examining the freshness, appearance, and overall condition of FV to ensure they meet the required standards. Improving collaboration and communication with suppliers will optimize supply chain operations and ensure that suppliers are aligned with the retailer‘s requirements. Corrective actions can be taken, and FV waste can be reduced.

Interviewed retailers acknowledged that *poor manipulation* can cause damage to the products, increasing the risk of spoilage (Fig. 3.26). One solution is that retailers provide their suppliers with dedicated transport crates for depositing the FV. This approach allows the products to be delivered directly to the shelves in these crates, minimizing additional handling and reducing the risk of damage. Of course, providing constant training for the employees involved in product handling is crucial. The training programs should focus on educating workers on proper handling techniques, storage procedures, and quality control measures.

Some of the causes mentioned above either originate from or are interconnected with another factor: the *high customer standards* regarding fresh products. This was

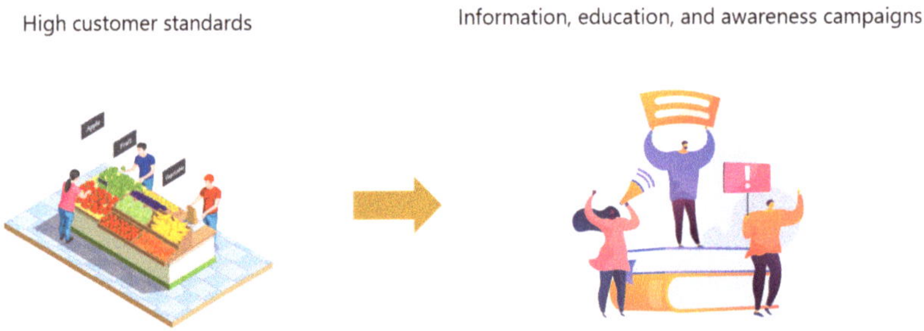

Fig. 3.27 Solution to reduce FV waste caused by high customer standards

a recurring cause encountered during the project interviews, and it was mentioned by members of various supply chain categories, including farmers, processors, retailers, and consumers (Fig. 3.27). FV with irregular forms or those of lower quality (although still suitable for consumption) face challenges in being sold, leading to eventual spoilage on the shelves. Consumers also expect a high variety of products and demand that they be available in abundance, putting pressure on retailers to meet their demands and, in turn, creating overstocks. In this case, effective communication strategies can be implemented, along with information, education, and awareness campaigns. These campaigns can focus on increasing awareness of sustainable consumption practices and the environmental impact of overstocking. Learning about the lifecycle of products and the importance of responsible purchasing behavior, consumers can make conscious decisions that benefit both the environment and the overall supply chain efficiency.

FV supply chain is a dynamic and interconnected space where the actions of one group affect the others, opening the way to both problems and solutions. Thus, certain actions in the retail sector could cause FV waste in other sectors. The strict aesthetic standards imposed by retailers may cause waste in the primary production sector. Marketing strategies, such as “buy one, get one free,” can lead to waste at the consumer level. These promotional approaches incentivize consumers to purchase quantities exceeding their actual needs, ultimately leading to surplus items that often end up in the trash. An option would be to offer discounts on single items (Infographic 3.5).

3.6 Consumers

According to Piras et al. (2022), consumers are responsible for 53% of the total food waste along the food supply chain in the EU. Interviews with consumers shed light on their perception regarding the main causes that generate FV waste in this sector.

The main causes mentioned by the interviewed consumers are the *lack of planning and excessive shopping* (Fig. 3.28). Consumers tend to buy more than they need, without having a shopping list or a plan for the meals they want to prepare in

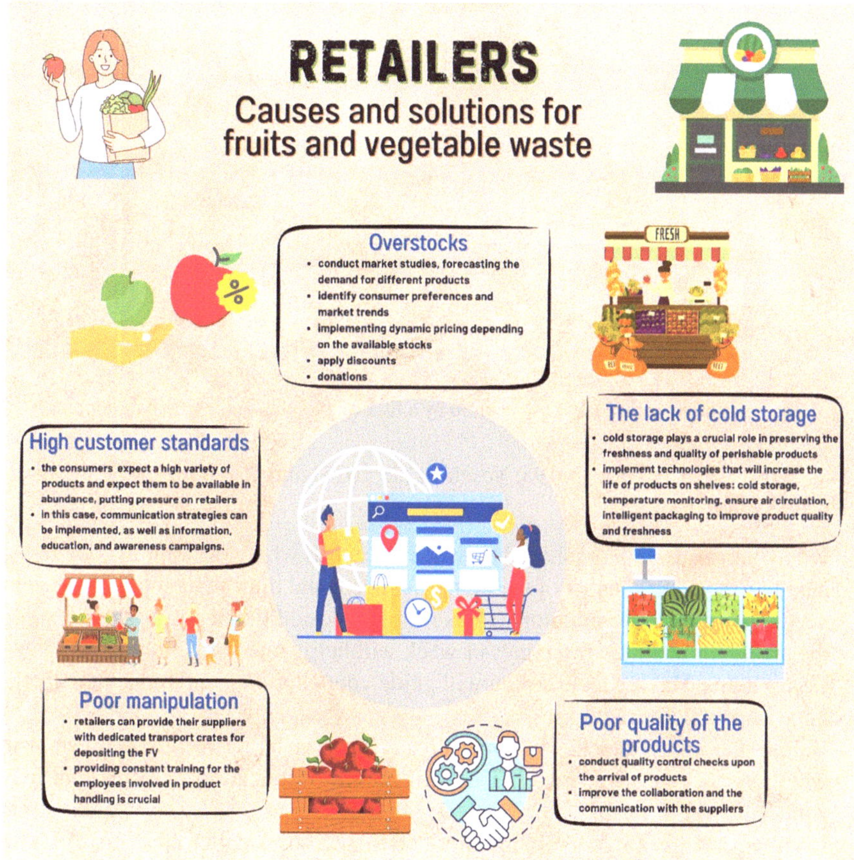

Infographic 3.5 Causes and solutions for FV waste in the retail sector

the next days. This leads to a surplus of perishable items that may not be consumed before their expiration date, ultimately resulting in avoidable waste. Impulse purchases of new products will also result in waste if the consumer doesn't like the products (Priefer et al., 2016). Moreover, according to STREFOWA (2019a), the promotions or other types of discounts encourage the consumers to buy more than they need or can consume.

One consumer commented:

> *The only reason is that we do not properly estimate how much we consume. We always buy more than we can consume within a given timeframe, and what is not consumed for an extended duration ends up spoiling; that is what leads to the situation where these fruits are wasted. So, we don't estimate correctly, we don't buy consciously, we aren't conscientious about shopping.*

Below are listed several guidelines and practical solutions that may help consumers avoid FV waste due to excessive shopping and overstocks (as suggested by Simões

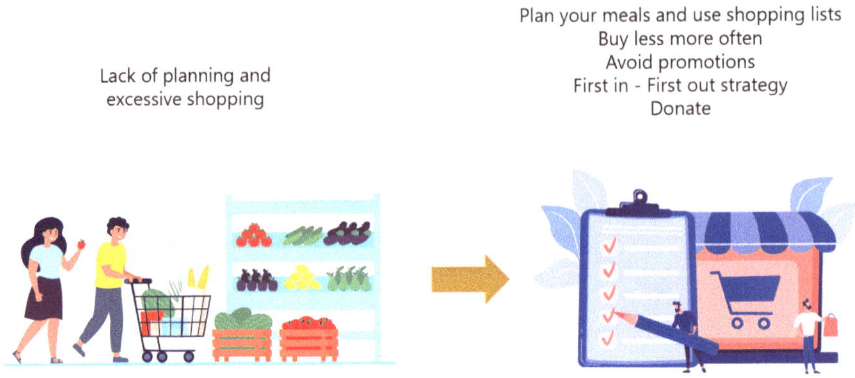

Fig. 3.28 Solution to reduce FV waste caused by a lack of planning and excessive shopping

et al., 2023; STREFOWA, 2019a; Veselá et al., 2023) and also by some of the interviewed consumers:

- Plan your meals and use shopping lists. Plan ahead what you want to cook in the next few days, make a list of the items you need, and limit your shopping to your needs, avoiding the temptation to make unplanned additional purchases. Planning the meals for the upcoming days or week will help you avoid purchasing unnecessary items. Having a clear plan will guide your shopping and reduce the likelihood of impulsive purchases.
- Before shopping, check what you already have at home in order to establish which items you need to buy. This will prevent duplication of items that may contribute to FV waste.
- It's always better to go shopping more often and buy smaller amounts of FV in order to avoid overstocks.
- Avoid promotions "buy one get one free" or other promotions that encourage bulk buying. Consider the shelf life of the items on promotion. If the quantity exceeds what you can reasonably consume before the products expire, it may lead to unnecessary waste. Moreover, the supermarkets often provide prepackaged FV that may contain more products than the consumer needs or include varieties unsuitable for the customer's preferences. Purchasing items you don't enjoy or won't use can lead to them being discarded, contributing to avoidable food waste.
- Make a habit of eating the most perishable FV first, applying also a "First in, First out" strategy. Organize the perishable items in your refrigerator in a way that allows for easy visibility and access.
- Use online platforms to find NGOs where you can donate FV. Reach out to food banks or shelters and create collaborations.

Poor storage management is another factor that will generate FV waste, according to the interviewed consumers (Fig. 3.29). The lack of knowledge regarding the proper storage of FV will accelerate the spoilage of these products. Therefore,

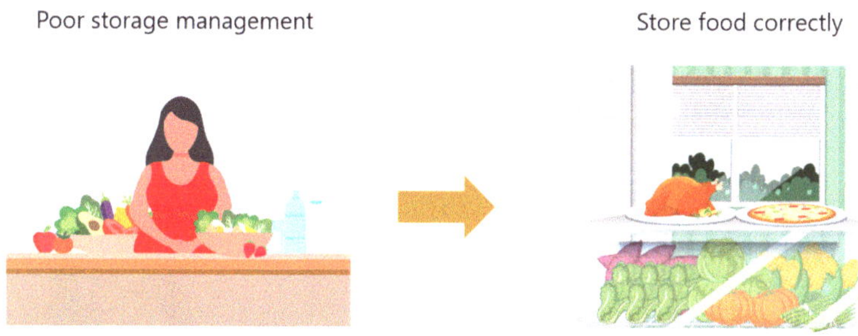

Fig. 3.29 Solution to reduce FV waste caused by poor storage management

consumers must provide optimal storage conditions and learn how to organize the products within the refrigerator.

The ideal temperature in the refrigerator is between 3 and 5 °C (Ghid de consum si pastrare a alimentelor, 2020). The last drawer in the refrigerator is intended for FV, providing the right temperature and controlled humidity. In this case, the FV should not be stored in plastic bags (Ghid de consum si pastrare a alimentelor, 2020).

Some consumers suggested that surplus vegetables that are not consumed immediately can be preserved by grating them, placing them in plastic bags, and then storing them in the freezer. This preservation method enables the preservation of vegetables for an extended period, making them readily available for future use in various culinary applications, such as soups, stews, and other dishes. Grating the vegetables before freezing facilitates convenient portioning and ensures they can be seamlessly incorporated into your cooking, contributing to a more efficient and sustainable approach to managing your vegetable supply.

Certain FV produce ethylene as they mature, a plant hormone that can accelerate fruit ripening. Understanding ethylene dynamics is essential for effective preservation of different FV. To prevent premature ripening or decay, it's advisable to keep ethylene-sensitive produce separate from items that emit ethylene (Ghid de consum si pastrare a alimentelor, 2020). Ethylene-producing items include bananas, apples, avocados, tomatoes, and melons. Ethylene-sensitive items include leafy greens, berries, carrots, and broccoli.

Interviewed consumers mentioned that *oversized meals* also cause FV waste. This can be related to the consumer‘s incorrect estimation of the amount of food they can eat at one meal, or it may refer to take-out orders. Estimating the amount of food required for a single-family meal can be challenging, especially for those without prior experience. Usually, people prefer to prepare more food than might be necessary to avoid the possibility of insufficient portions. This results in a food surplus that could go to waste. Therefore, this cause is related to another one regarding the *lack of skills for food preparation* (Fig. 3.30). Poor planning or cooking experience can lead to an excess of food, as well as cooking mistakes that generate waste.

To address the issue of oversized meals and reduce food waste, one solution is to learn from experience. This involves paying more attention to the food that is

Fig. 3.30 Solution to reduce FV waste caused by oversized meals and a lack of skills

actually needed for one meal, the amounts of leftovers, and then adjusting the amount of food that is cooked based on this information. For instance, noting the leftover vegetables after a meal can help adjust the quantity cooked next time to match the family's appetite more accurately. Another solution is to learn how to recombine leftovers into new and creative meals. For instance, the vegetables left over from the side dishes can be used to cook creamy soups, or they can be fried and used as fillings for pies or cakes (Ghid de consum si pastrare a alimentelor, 2020). This approach not only minimizes waste but also adds variety to subsequent meals. Consumers may also consider new methods to utilize the FV that have lost their quality, but are still suitable for consumption. This FV can be used to make smoothies, milkshakes, and vegetable cream soups.

The poor quality of FV may also influence the quantities of FV waste generated by consumers (Fig. 3.31). If the products don't meet the customer's expectations, they will not be consumed and will end up in the trash. The taste is the most important factor that determines whether the consumer will throw away the FV instead of consuming it. When FV experience uneven ripening, resulting in some sections being excessively ripe while others remain underdeveloped, consumers might discard the entire item due to the inconvenience of separating the usable parts from those that have deteriorated.

To avoid these problems, some consumers mentioned that they thoroughly inspect FV before purchasing, checking for any visible blemishes, signs of decay, or uneven ripening. Choosing products carefully can prevent dissatisfaction and subsequent waste. Buying the smaller quantities can be useful to ensure that the FV are consumed before they lose freshness. Consumers should also consider using products that may not have the desired flavor or sweetness in new and innovative ways. For instance, you can combine less sweet fruits or those that are lightly overripe with fruits that possess the desired overall taste, creating smoothies or milkshakes.

Interviewed consumers draw attention to the fact that a *lack of education and interest in reducing food waste* is another cause that must be addressed (Fig. 3.32). Many individuals may not fully understand the environmental, economic, and social implications of food waste. Limited awareness can lead to a lack of motivation to

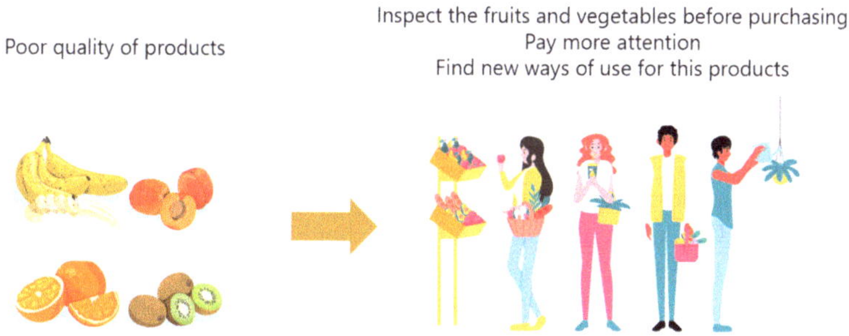

Fig. 3.31 Solution to reduce FV waste caused by the poor quality of the products

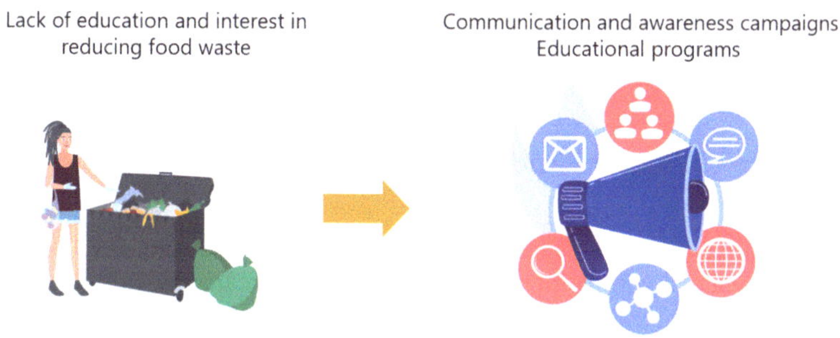

Fig. 3.32 Solution to reduce FV waste caused by a lack of education and interest in reducing food waste

change behaviors. Moreover, people may perceive reducing food waste as inconvenient and time-consuming. The effort required to adopt new habits may deter individuals from making changes.

One consumer expressed the need for a mentality shift to reduce food waste:

> *Well, first of all, the world should be educated in this sense. Let's say a workshop on how to avoid food waste, or how to reuse it, and so on. It just takes a new mentality and a whole education about this thing, so that everyone learns to do this and not consume, not buy more than they need, and things like that. So, we need to change our mentality somehow.*

Another consumer explained:

> *We need more education because I really believe that we do not receive the proper education for this problem, for waste, neither in schools nor at home. And why not educate yourself? There is the Internet, social networks, and a plethora of information. Thus, you can inform yourself, especially if we are talking about adults.*

According to Piras et al. (2022), an efficient method to increase people's interest in food waste is the communication campaigns with a focus on practical goals, showing that food reduction has positive impacts not only on environmental protection but can also reduce household expenses. This can include practical guidance on how to integrate waste reduction practices into everyday life—for instance, tips on meal planning, understanding expiration dates, utilizing leftovers creatively, and being mindful of portion sizes during meal preparation. Moreover, more attention should be given to educational programs at various levels, including schools, community centers, and online platforms. These initiatives aim to raise awareness about the environmental impact of food waste, highlight the economic consequences, and offer practical tips on minimizing waste through proper storage, meal planning, and repurposing leftovers. Awareness campaigns should be tailored to different group types, reaching individuals at various stages of life. Initiating sustainable habits at an early stage, such as during the school years, is crucial in instilling a lasting sense of responsibility toward the environment. Children can be engaged through interactive and age-appropriate activities that not only educate them about the impact of food waste but also empower them to make informed choices (Aschemann-Witzel et al., 2017; Begho & Fadare, 2023; European Commission. Joint Research Centre, 2023b) (Infographic 3.6).

3.7 Waste Collectors and Processors

Waste collectors and processors play a crucial role in reducing food waste and recovering its value by implementing sustainable waste management practices, promoting awareness, and collaborating with various stakeholders. Additionally, they can play an educational role by promoting sustainable waste management practices; informing households, businesses, and communities about the environmental impact of food waste; and encouraging the adoption of strategies to minimize waste generation. The actions of waste collectors can contribute to food waste prevention and reduction, facilitating the recovery of these resources through composting or anaerobic digestion processes.

Interviewed waste collectors stated that a problem preventing a greater reduction of food waste in many regions is the *lack of composting facilities* (Fig. 3.33). This issue poses a critical challenge, as without proper infrastructure for composting, a substantial portion of food waste that could be repurposed as compost ends up being deposited in landfills.

A waste collector pointed to the need to improve the waste collection and processing infrastructure at the country level:

> *I think that the lack of a recovery/composting system at the county level, at a high level, a treatment station, which can take the large amount of organic waste that exists. So, this is my opinion. I don't think it would be that complicated to collect them separately. There are companies that do this.*

Therefore, an effective solution would be to invest and allocate resources in a comprehensive composting infrastructure. This strategic intervention aims to tackle the core issue by creating the necessary facilities to process and repurpose organic waste effectively.

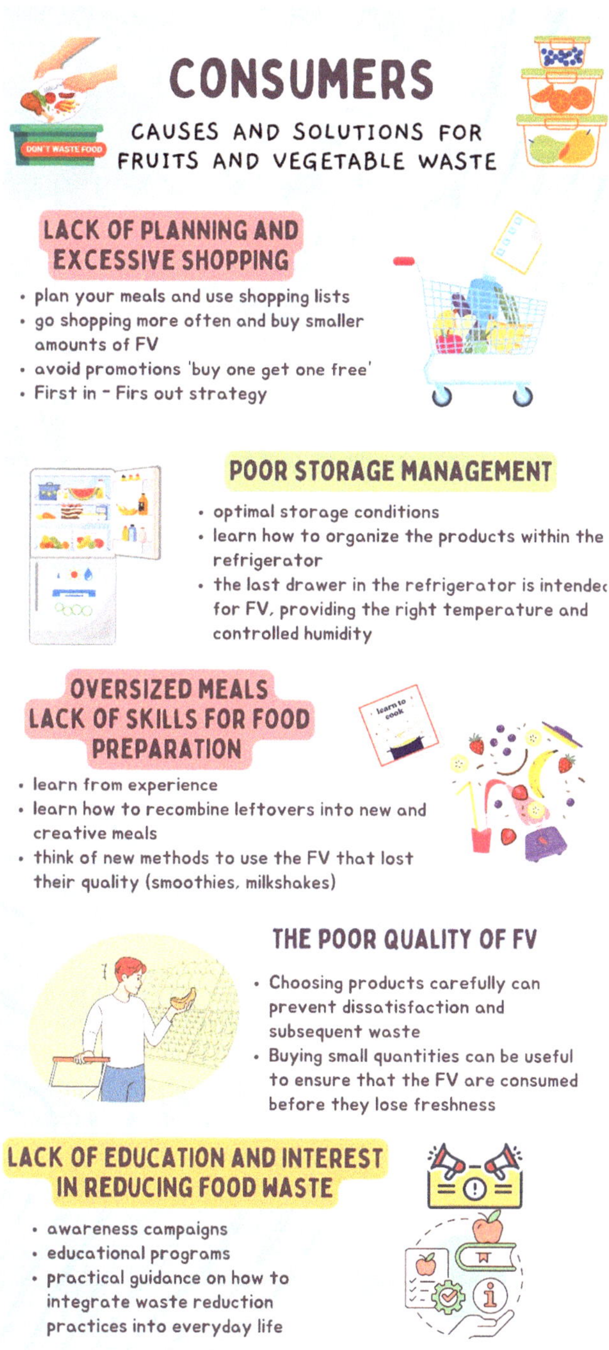

Infographic 3.6 Causes and solutions for FV waste in the consumer sector

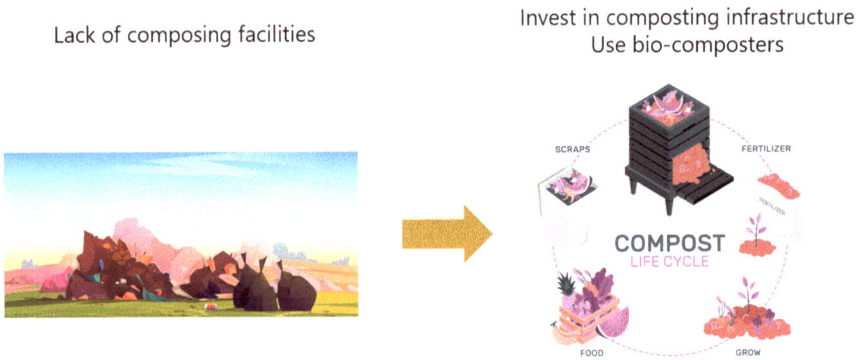

Fig. 3.33 Solution to reduce FV waste caused by a lack of composting facilities

Another solution mentioned by the interviewed waste collectors is the use of bio-composters. Bio-composters represent an innovative and efficient solution for managing organic matter, significantly contributing to the reduction of biodegradable waste. These devices have a specially designed structure to facilitate the composting process and transform organic matter into compost without requiring complex infrastructure. They can be used at both individual and community levels, in households or apartments, as well as in larger communities.

By supporting and encouraging the large public and various stakeholders to actively participate in food waste reduction efforts, a transformative shift in behavior can be achieved. For instance, different stakeholders can compost their FV (composting at the source).

One waste collector recommended:

> *As a retailer, you can return fruits and vegetables to your suppliers for composting. You can establish an agreement with them to receive some benefits for the resulting compost. If you are a producer, you can turn the losses into compost and use them as fertilizer for what you produce.*

The *lack of a correct or effective selective collection system* is another significant issue contributing to the generation of food waste, according to waste collectors (Fig. 3.34). This can be related to the inadequate infrastructure for selective waste collection, lack of public awareness of problems with the implementation (inadequate enforcement, insufficient monitoring, or a lack of standardized practices across different regions).

One waste collector recommended:

> *As for organic waste, the solution exists, but we need to have the conditions as well. And this means a platform, a correct selective collection system, an intelligent and appropriate management system, so that we will no longer talk about waste, we will talk about recovery.*

Therefore, the waste collection systems should be improved to increase their efficiency and ensure convenient access for the population. This system should be reliable, flexible, easy to use, and require minimal effort from the population. The bins and containers should be strategically placed in easily accessible locations, within

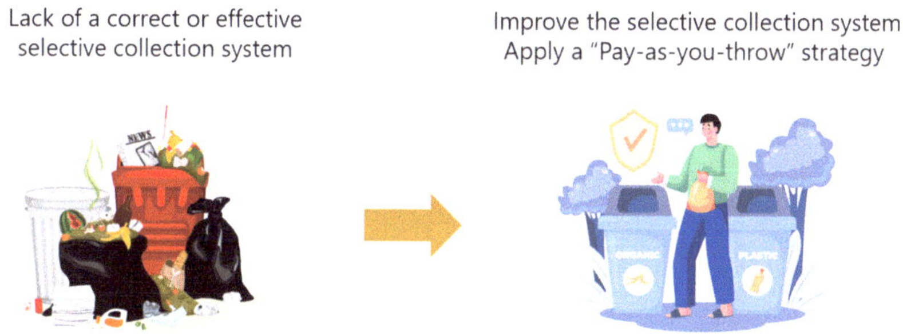

Fig. 3.34 Solution to reduce FV waste caused by a lack of a selective collection system

short walking distances. Optimal placement ensures that individuals, whether in residential or commercial areas, can conveniently dispose of their waste, minimizing the effort required for proper waste disposal. Moreover, according to STREFOWA (2019d), constant communication with the community will be useful in gathering feedback, improving the system, and addressing potential problems.

Another solution is to apply a "Pay-as-you-throw" strategy wherein households are billed based on the quantity of waste they produce (STREFOWA, 2019b). Interview participants repeatedly suggested this solution. By directly linking costs to the volume of waste generated, this strategy encourages households to adopt waste reduction practices. Residents become more inclined to engage in recycling, composting, and overall mindful consumption patterns, thereby optimizing their waste output and subsequently lowering disposal costs.

The waste collectors also highlighted that the *educational level of the population* regarding food waste and the correct way to collect and recycle is another issue that must be addressed (Fig. 3.35). Communication channels should be established to inform and motivate the population to reduce waste generation, while also promoting proper collection and recycling practices. More information should be provided regarding the system's infrastructure, guidelines for waste sorting, designated collection times, and the system's outcomes. This way, the public is well informed and engaged in the entire waste management process. Another effective strategy is to implement educational programs tailored to different demographics, emphasizing the consequences of food waste and the significance of proper waste management. These programs should not only inform but also inspire behavioral changes (Manika et al., 2022).

Of course, efforts to reduce food waste in this sector should extend beyond communication and providing information to the general population, and should encompass a comprehensive strategy that facilitates information exchange, collaboration, and the sharing of best practices among all relevant stakeholders. This approach recognizes the interconnected nature of the food supply chain and emphasizes the importance of engaging various entities involved in waste management (Infographic 3.7).

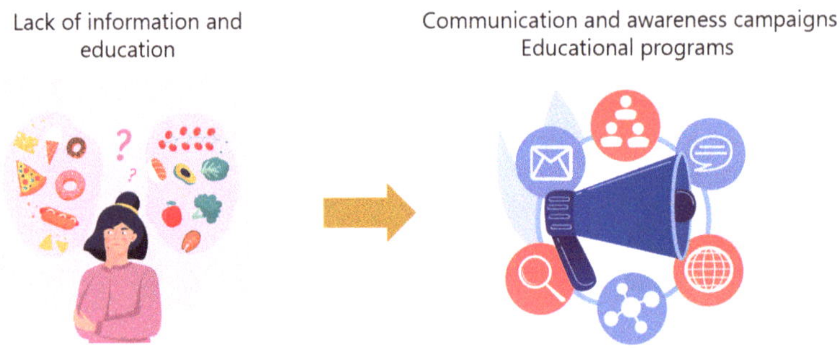

Fig. 3.35 Solution to reduce FV waste caused by a lack of information and education

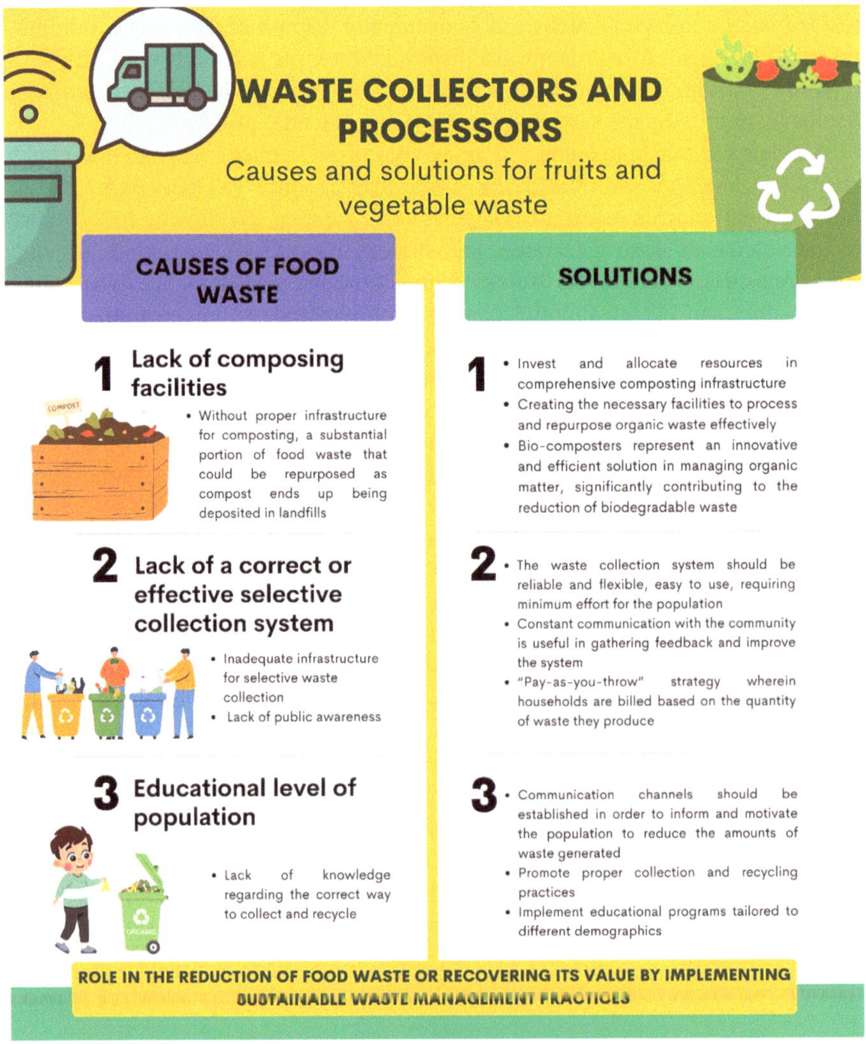

Infographic 3.7 Causes and solutions for FV waste in the waste collectors sector

3.8 Policymakers

Policymakers play a crucial role in driving systemic change and creating legal frameworks that address the problem of food waste (Fig. 3.36). Their decisions and initiatives can shape regulations, incentives, and frameworks that influence behaviors across the entire food supply chain. According to Ada et al. (2021), some of the issues that must be addressed at this level, because they represent barriers to a circular economy, are related to the absence of suitable legal frameworks, policy complexities, taxation and incentive issues, lenient environmental regulations, divergent priorities between central and local governments, and inadequate waste infrastructure.

Policymakers have the responsibility to enact and enforce legislation, design and implement incentive programs, initiate public education campaigns to raise awareness about environmental issues related to food waste, and align food waste reduction policies with broader sustainability goals.

Several stakeholders interviewed in the "Team Up for Less Food Waste" project highlight a critical concern: the absence of a comprehensive legal framework to facilitate the donation process. They stated that the process can be quite complicated and, therefore, difficult in terms of legislation and the obligations they have,

Fig. 3.36 Solution to reduce FV waste in the policymakers sector

especially for retailers. It may also be expensive in certain instances. It is cheaper to throw the FV away than to donate them. Therefore, a clear and comprehensive legal framework may create a foundation for addressing these issues, providing some basic guidelines and simplifying the entire process of food donation, making it more accessible, cost-effective, and aligned with regulatory obligations.

Policymakers should initiate more public education campaigns to raise awareness about the environmental, economic, and social impacts of food waste. Policy makers can use their platforms to disseminate information and encourage responsible consumer behavior, proper food storage, and waste reduction practices (Bos-Brouwers et al., 2020; European Commission. Joint Research Centre, 2023b; Shen et al., 2023).

One policymaker commented:

> *Education: We have many programs with the school inspectorate. Normally, we start education at an early age, where even through this program called ‹‹milk, croissant, and apple››, the child is taught to respect that material from an early age. Additionally, there are alternative schooling systems, alongside the mainstream ones, where students are taught to respect nature and the work involved in obtaining those products. Apart from that, obviously, the so-called "seven years from home" should exist, where all the parents are obliged to educate their children in the direction of non-generation of waste.*

Moreover, according to the interviewed policymakers, more emphasis should be placed on national strategies for promoting the quality of local products, which also includes promoting short supply chains. These strategies should include financial and logistical support to local producers to enhance their capacity to meet quality standards. This support can take various forms, including grants, training programs, and access to resources that enable farmers and manufacturers to optimize their production processes. On the other hand, the development of short supply chains ensures a direct link between producers and consumers. This not only reduces the distance between production and consumption but also fosters transparency, traceability, and a stronger connection between local communities and the sources of their food. Considering this direction, interviewed farmers requested support from policymakers to facilitate their access to supermarkets. For example, farmers suggested having a regulation that requires retailers to prioritize local farmers or even impose on them to include local farmers among their FV suppliers.

Policymakers acknowledged their responsibility to support sustainable practices, proposing funding schemes for the development of composting systems that contribute to reducing food waste. Another purpose of this funding scheme could be the development of common storage facilities that can serve as centralized hubs where farmers, particularly those with limited financial resources, can bring their products. This collective approach not only reduces individual investment burdens but also creates shared infrastructure that optimizes resource utilization. By offering financial incentives for farmers to utilize such facilities, policymakers can create a collaborative environment, fostering community-driven solutions to reduce food waste.

To support sustainable and environmentally friendly practices among stakeholders, a comprehensive system of incentives should be introduced. This approach can include a combination of financial benefits and regulatory incentives, aiming to encourage a meaningful shift toward responsible practices. Several examples of such strategies include establishing tax incentives, introducing subsidies or financial support programs for local producers who demonstrate a commitment to sustainable practices, providing preferential treatment in regulatory processes, and facilitating market access. Moreover, to discourage excessive waste generation, an additional strategy involves applying fees for waste disposal. This approach aligns with the "polluter pays" principle, where those generating waste bear the associated costs. By implementing fees based on the volume or type of waste produced, individuals and businesses are incentivized to reduce waste at the source, implement recycling practices, and adopt more sustainable consumption habits (European Commission. Joint Research Centre., 2023b; Shen et al., 2023).

A policymaker gave details about a successful solution applied in the past:

> *Compensatory payments or subsidies: these are the ways in which the ministry can influence both the production quantity and the sales chain. For example, we used to have a form of compensatory payments for tomatoes, and as a result, a very large quantity of tomatoes was generated in the country, which, of course, was planned from the beginning. Farmers knew that they had a compensatory payment for the tomatoes that were processed. We generated a very favorable price for the final beneficiary, the consumer of tomato products, and, of course, also for the companies involved in this part of the chain. Overall, everyone was happy.*

Another action policymakers can take is to facilitate partnerships with businesses, non-profit organizations, and communities to develop collaborative strategies for reducing food waste. Engaging stakeholders from various sectors ensures a holistic and inclusive approach to addressing the complexities of food waste (Infographic 3.8).

3.9 NGOs

NGOs can have a significant impact on reducing food waste through their actions (Fig. 3.37).

NGOs can support and facilitate the donation process by engaging in direct food recovery and distribution efforts. They collaborate with various stakeholders in the food supply chain to collect FV that are in surplus and redistribute them to food banks, shelters, and communities in need. They can also support stakeholders in implementing sustainable practices and waste reduction strategies. This may include recommendations for sustainable actions, environmentally friendly production processes, and the incorporation of renewable resources. NGOs can serve as knowledge hubs, sharing best practices that align with broader sustainability goals (EU Platform on Food Losses and Food Waste, 2019; Fattibene et al., 2020).

Another role of NGOs is to participate in educational initiatives, raising awareness about the environmental, social, and economic impacts of food waste. Educational initiatives conducted by NGOs target various stakeholders, including businesses, consumers, and policymakers to promote understanding and behavioral change. Within these actions, they offer practical and easy-to-implement strategies,

Infographic 3.8 Solution to reduce FV waste in the policymakers sector

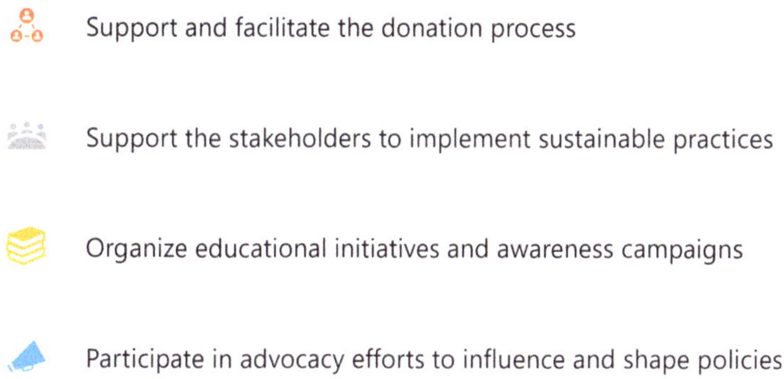

Fig. 3.37 The role of NGOs in food waste reduction

providing useful information such as meal planning or proper storage practices that can lead to significant waste reduction without imposing a major inconvenience. Moreover, they organize awareness campaigns, engaging the community in food waste reduction initiatives (European Commission. Joint Research Centre, 2023b; Fattibene et al., 2020).

An NGO representative explained:

> *We have a list of solutions ... So, we recommend everyone to plan their meals, use everything they bought. We address the management aspect to reduce the size of the dishes and to appreciate less attractive fruits and vegetables. One of everyone's favorite workshops is to make pasta with imperfect vegetables. That is, we bring the ugliest vegetables that we find in the market (in supermarkets, we don't really find that kind of product) and we trim them. The bio-waste resulting from this process is used for composting. We put the trimmed vegetables in the compost bin and blend them with a little water. Essentially, that's the wet part of the flour. So, flour plus this vegetable smoothie (because it's only vegetables), and we make rainbow pasta. So, there are all the colors: beet, one color, carrot, another color. And so, we demonstrate that even ugly fruits and vegetables have nutritional value and can be utilized as such without being rejected or discarded. Following that, there are guidelines for storage and preservation in the pantry, refrigerator, and freezer. We give them recipes to cook with leftovers, all these pies are named quiche, because they have borrowed the name. Another campaign that we have is to take away the ordered food or the leftovers, so as not to leave it on the plate. And even if I don't know, it goes to a street person or an animal, at least it doesn't end up in the trash.*

NGOs actively participate in advocacy efforts to influence and shape policies related to food waste reduction. They work with governments and international bodies to advocate for supportive regulations, incentives for waste reduction, and the development of frameworks that facilitate food donation and recycling (Infographic 3.9).

Infographic 3.9 The role of NGOs in food waste reduction

References

Ada, N., Kazancoglu, Y., Sezer, M. D., Ede-Senturk, C., Ozer, I., & Ram, M. (2021). Analyzing barriers of circular food supply chains and proposing industry 4.0 solutions. *Sustainability, 13*(12), 6812. https://doi.org/10.3390/su13126812

Al-Dairi, M., Pathare, P. B., Al-Yahyai, R., & Opara, U. L. (2022). Mechanical damage of fresh produce in postharvest transportation: Current status and future prospects. *Trends in Food Science & Technology, 124*, 195–207. https://doi.org/10.1016/j.tifs.2022.04.018

Aragie, E., Balié, J., & MoralesOpazo, C. (2018). Does reducing food losses and wastes in sub-Saharan Africa make economic sense? *Waste Management & Research: The Journal for a Sustainable Circular Economy, 36*(6), 483–494. https://doi.org/10.1177/0734242X18770247

Archip, B. C., Banatean-Dunea, I., Petrescu, D. C., & Petrescu-Mag, R. M. (2023). Determinants of food waste in Cluj-Napoca (Romania): A community-based system dynamics approach. *International Journal of Environmental Research and Public Health, 20*(3), 2140. https://doi.org/10.3390/ijerph20032140

Aschemann-Witzel, J., De Hooge, I. E., Rohm, H., Normann, A., Bossle, M. B., Grønhøj, A., & Oostindjer, M. (2017). Key characteristics and success factors of supply chain initiatives tackling consumer-related food waste—A multiple case study. *Journal of Cleaner Production, 155*, 33–45. https://doi.org/10.1016/j.jclepro.2016.11.173

Bajželj, B., Quested, T. E., Röös, E., & Swannell, R. P. J. (2020). The role of reducing food waste for resilient food systems. *Ecosystem Services, 45*, 101140. https://doi.org/10.1016/j.ecoser.2020.101140

Begho, T., & Fadare, O. (2023). Does household food waste prevention and reduction depend on bundled motivation and food management practices? *Cleaner and Responsible Consumption, 11*, 100142. https://doi.org/10.1016/j.clrc.2023.100142

Boiteau, J. M., & Pingali, P. (2023). Can we agree on a food loss and waste definition? An assessment of definitional elements for a globally applicable framework. *Global Food Security, 37*, 100677. https://doi.org/10.1016/j.gfs.2023.100677

Bos-Brouwers, H. E. J., Burgos, S., Colin, F., Graf, V., van Herpen, H. W. I., van Trijp, J. C. M., & van Geffen, E. J. (2020). Policy recommendations to improve food waste prevention and valorisation in the EU: Deliverable D 3.5. Refresh.

Calderón-Oliver, M., & López-Hernández, L. H. (2022). Food vegetable and fruit waste used in meat products. *Food Reviews International, 38*(4), 628–654. https://doi.org/10.1080/87559129.2020.1740732

I. Ajtai et al. (eds.), *Guidelines for Food Waste Prevention and Reduction: Insights from the Fruit and Vegetable Sector*, https://doi.org/10.1007/978-3-032-14506-2

Cane, M., & Parra, C. (2020). Digital platforms: Mapping the territory of new technologies to fight food waste. *British Food Journal, 122*(5), 1647–1669. https://doi.org/10.1108/BFJ-06-2019-0391

Chen, C., Chaudhary, A., & Mathys, A. (2020). Nutritional and environmental losses embedded in global food waste. *Resources, Conservation and Recycling, 160*, 104912. https://doi.org/10.1016/j.resconrec.2020.104912

Craswell, E. (2021). Fertilizers and nitrate pollution of surface and ground water: An increasingly pervasive global problem. *SN Applied Sciences, 3*(4), 518.

De Souza, R. M., Seibert, D., Quesada, H. B., De Jesus Bassetti, F., Fagundes-Klen, M. R., & Bergamasco, R. (2020). Occurrence, impacts and general aspects of pesticides in surface water: A review. *Process Safety and Environmental Protection, 135*, 22–37. https://doi.org/10.1016/j.psep.2019.12.035

Eičaitė, O., Baležentis, T., Ribašauskienė, E., Morkūnas, M., Melnikienė, R., & Štreimikienė, D. (2022). Food waste in the retail sector: A survey-based evidence from central and Eastern Europe. *Journal of Retailing and Consumer Services, 69*, 103116. https://doi.org/10.1016/j.jretconser.2022.103116

EU Platform on Food Losses and Food Waste. (2019). *Redistribution of surplus food: Examples of practices in the Member States* (p. 89). EU Platform on Food Losses and Food Waste. https://food.ec.europa.eu/system/files/2019-06/fw_eu-actions_food-donation_ms-practices-food-redis.pdf

European Commission. (n.d.). *Waste framework directive*. https://environment.ec.europa.eu/topics/waste-and-recycling/waste-framework-directive_en

European Commission. (2021). *Brief on food waste in the European Union*. https://food.ec.europa.eu/system/files/2021-04/fw_lib_stud-rep-pol_ec-know-cen_bioeconomy_2021.pdf

European Commission. Joint Research Centre. (2023a). *Assessing the economic, social and environmental impacts of food waste reduction targets: A model based analysis*. Publications Office. https://data.europa.eu/doi/10.2760/77251

European Commission. Joint Research Centre. (2023b). *Tools, best practices and recommendations to reduce consumer food waste: A compendium*. Publications Office. https://data.europa.eu/doi/10.2760/967005

European Environment Agency. (2023). *Waste prevention country profile—Romania* (p. 16). https://www.eea.europa.eu/themes/waste/waste-prevention/countries/2023-waste-prevention-country-fact-sheets/romania_waste_prevention_2023

European Parliament & Council of Ministers. (2002). *Regulation no. 178/2002 laying down the general principles and require ments of food law, establishing the European Food SafetyAuthority and laying down procedures in matters of food safety*. Official Journal of the European Communities L 31/1. https://eur-lex.europa.eu/legal-content/EN/TXT/PDF/?uri=CELEX:32002R0178

European Parliament & Council of Ministers. (2008). *Directive 2008/98/EC on waste and repealing certain Directives*. Official Journal of the European Union L 312/3. https://eur-lex.europa.eu/legal-content/EN/TXT/HTML/?uri=CELEX:32008L0098

European Parliament & Council of Ministers. (2018). *Directive (EU) 2018/851 amending Directive 2008/98/EC on waste*. Official Journal of the European Union L 150/109. https://eur-lex.europa.eu/eli/dir/2018/851/oj?locale=en

EUROSTAT. (2024a). *Crop production in EU standard humidity* [dataset]. EUROSTAT. https://ec.europa.eu/eurostat/databrowser/view/APRO_CPSH1__custom_10004870/default/table?lang=en

EUROSTAT. (2024b). *Crop production in EU standard humidity* [dataset]. EUROSTAT. https://ec.europa.eu/eurostat/databrowser/view/APRO_CPSH1__custom_10004834/default/table?lang=en

EUROSTAT. (2024c). *Final consumption expenditure of households by consumption purpose (COICOP 3 digit)* [dataset]. EUROSTAT. https://ec.europa.eu/eurostat/databrowser/view/nama_10_co3_p3__custom_10700708/default/table?lang=en

EUROSTAT. (2024d). *Final consumption expenditure of households by consumption purpose (COICOP 3 digit)* [dataset]. https://ec.europa.eu/eurostat/databrowser/view/nama_10_co3_p3__custom_10760089/default/table?lang=en

EUROSTAT. (2024e). *Fruit and vegetable production in 2022* [dataset]. EUROSTAT. https://ec.europa.eu/eurostat/web/products-eurostat-news/w/ddn-20240301-1#:~:text=The%20EU%20produces%20a%20widetonnes%20of%20crops%20were%20harvested

Falcon, W. P., Naylor, R. L., & Shankar, N. D. (2022). Rethinking global food demand for 2050. *Population and Development Review, 48*(4), 921–957. https://doi.org/10.1111/padr.12508

FAO. (2013). *Food wastage footprint: Impacts on natural resources*. https://www.fao.org/3/i3347e/i3347e.pdf

FAO. (2015). *Food wastage footprint & climate change*. https://openknowledge.fao.org/server/api/core/bitstreams/7fffcaf9-91b2-4b7b-bceb-3712c8cb34e6/content

FAO. (2019). *The state of food and agriculture 2019: Moving forward on food loss and waste reduction*. UN.

FAO. (2022). *The state of the world's land and water resources for food and agriculture 2021—Systems at breaking point*. FAO. https://doi.org/10.4060/cb9910en

FAO. (2023). *Sustainability pathways. Food wastage footprint*. https://www.fao.org/nr/sustainability/food-loss-and-waste/en/

Fattibene, D., Recanati, F., Dembska, K., & Antonelli, M. (2020). Urban food waste: A framework to analyse policies and initiatives. *Resources, 9*(9), 99. https://doi.org/10.3390/resources9090099

FUSIONS. (2016). *Food Waste Definition*. https://www.eu-fusions.org/index.php/about-food-waste/280-food-waste-definition

Ganesh, K. S., Sridhar, A., & Vishali, S. (2022). Utilization of fruit and vegetable waste to produce value-added products: Conventional utilization and emerging opportunities-a review. *Chemosphere, 287*, 132221. https://doi.org/10.1016/j.chemosphere.2021.132221

Gardas, B. B., Raut, R. D., & Narkhede, B. (2018). Evaluating critical causal factors for post-harvest losses (PHL) in the fruit and vegetables supply chain in India using the DEMATEL approach. *Journal of Cleaner Production, 199*, 47–61. https://doi.org/10.1016/j.jclepro.2018.07.153

Ghid de consum si pastrare a alimentelor. (2020). https://foodwastecombat.com/wp-content/uploads/2021/05/Ghidul-anti-risipa-Food-Waste-Combat-2020-Web.pdf

Göbel, C., Langen, N., Blumenthal, A., Teitscheid, P., & Ritter, G. (2015). Cutting food waste through cooperation along the food supply chain. *Sustainability, 7*(2), 1429–1445. https://doi.org/10.3390/su7021429

Gokarn, S., & Choudhary, A. (2021). Modeling the key factors influencing the reduction of food loss and waste in fresh produce supply chains. *Journal of Environmental Management, 294*, 113063. https://doi.org/10.1016/j.jenvman.2021.113063

Grizzetti, B., Pretato, U., Lassaletta, L., Billen, G., & Garnier, J. (2013). The contribution of food waste to global and European nitrogen pollution. *Environmental Science & Policy, 33*, 186–195. https://doi.org/10.1016/j.envsci.2013.05.013

Gustavsson, J., Cederberg, C., Sonesson, U., Van Otterdijk, R., & Meybeck, A. (2011). *Global food losses and food waste*. FAO.

Hamilton, S. F., Richards, T. J., Shafran, A. P., & Vasilaky, K. N. (2022). Farm labor productivity and the impact of mechanization. *American Journal of Agricultural Economics, 104*(4), 1435–1459. https://doi.org/10.1111/ajae.12273

Heikkilä, L., Reinikainen, A., Katajajuuri, J.-M., Silvennoinen, K., & Hartikainen, H. (2016). Elements affecting food waste in the food service sector. *Waste Management, 56*, 446–453. https://doi.org/10.1016/j.wasman.2016.06.019

HLPE. (2014). Food losses and waste in the context of sustainable food systems. *A report by the High Level Panel of Experts on Food Security and Nutrition of the Committee on World Food Security*. https://www.fao.org/3/i3901e/i3901e.pdf

Ingrao, C., Strippoli, R., Lagioia, G., & Huisingh, D. (2023). Water scarcity in agriculture: An overview of causes, impacts and approaches for reducing the risks. *Heliyon, 9*(8), e18507. https://doi.org/10.1016/j.heliyon.2023.e18507

IRCEM. (2021). *Study of food waste in Romania.*

Louis, D., & Lombart, C. (2018). Retailers' communication on ugly fruits and vegetables: What are consumers' perceptions? *Journal of Retailing and Consumer Services, 41*, 256–271. https://doi.org/10.1016/j.jretconser.2018.01.006

Manika, D., Iacovidou, E., Canhoto, A., Pei, E., & Mach, K. (2022). Capabilities, opportunities and motivations that drive food waste disposal practices: A case study of young adults in England. *Journal of Cleaner Production, 370*, 133449. https://doi.org/10.1016/j.jclepro.2022.133449

Mbow, C., Rosenzweig, C., Barioni, L. G., Benton, T. G., Herrero, M., Krishnapillai, M., Liwenga, E., Pradhan, P., Rivera-Ferre, M. G., Sapkota, T., Tubiello, F. N., & Xu. (2019). *Food security* (climate change and land: An IPCC special report on climate change, Desertification, Land Degradation, Sustainable Land Management, Food Security, and Greenhouse Gas Fluxes in Terrestrial Ecosystems, p. 114).

Mena, C., Adenso-Diaz, B., & Yurt, O. (2011). The causes of food waste in the supplier–retailer interface: Evidences from the UK and Spain. *Resources, Conservation and Recycling, 55*(6), 648–658. https://doi.org/10.1016/j.resconrec.2010.09.006

Milford, A. B., Lien, G., & Reed, M. (2021). Different sales channels for different farmers: Local and mainstream marketing of organic fruits and vegetables in Norway. *Journal of Rural Studies, 88*, 279–288. https://doi.org/10.1016/j.jrurstud.2021.08.018

Obersteiner, G., & Scherhaufer, S. (2021). Environmental impact of food waste. In *Environmental impact of agro-food industry and food consumption* (pp. 261–283). Elsevier. https://doi.org/10.1016/B978-0-12-821363-6.00011-4

Patel, S. (2012). Potential of fruit and vegetable wastes as novel biosorbents: Summarizing the recent studies. *Reviews in Environmental Science and Bio/Technology, 11*, 365–380.

Petrescu-Mag, R. M., Petrescu, D. C., Ajtai, I., Roba, C. A., Gica, O. A., Cuibus, L., Tenter, A., Toader, V., Negrusa, A. L., & Bican-Brişan, N. (2024). Causes and solutions for fruit and vegetable waste: A participatory approach with Romanian farmers for sustainable agriculture. *International Journal of Agricultural Sustainability, 22*(1), 2329391. https://doi.org/10.1080/14735903.2024.2329391

Piras, S., Righi, S., Setti, M., Koseoglu, N., Grainger, M. J., Stewart, G. B., & Vittuari, M. (2022). From social interactions to private environmental behaviours: The case of consumer food waste. *Resources, Conservation and Recycling, 176*, 105952. https://doi.org/10.1016/j.resconrec.2021.105952

Priefer, C., Jörissen, J., & Bräutigam, K.-R. (2016). Food waste prevention in Europe—A cause-driven approach to identify the most relevant leverage points for action. *Resources, Conservation and Recycling, 109*, 155–165. https://doi.org/10.1016/j.resconrec.2016.03.004

Read, Q. D., Brown, S., Cuéllar, A. D., Finn, S. M., Gephart, J. A., Marston, L. T., Meyer, E., Weitz, K. A., & Muth, M. K. (2020). Assessing the environmental impacts of halving food loss and waste along the food supply chain. *Science of the Total Environment, 712*, 136255. https://doi.org/10.1016/j.scitotenv.2019.136255

Rezaei, M., & Liu, B. (2017). *Food loss and waste in the food supply chain* (pp. 26–27). International Nut and Dried Fruit Council.

Romanian Parliament. (2016). Law No. 217/ 2016 regarding reduction of food waste. https://legislatie.just.ro/public/DetaliiDocument/183792

Sagar, N. A., Pareek, S., Sharma, S., Yahia, E. M., & Lobo, M. G. (2018). Fruit and vegetable waste: Bioactive compounds, their extraction, and possible utilization. *Comprehensive Reviews in Food Science and Food Safety, 17*(3), 512–531. https://doi.org/10.1111/1541-4337.12330

Scherhaufer, S., Moates, G., Hartikainen, H., Waldron, K., & Obersteiner, G. (2018). Environmental impacts of food waste in Europe. *Waste Management, 77*, 98–113. https://doi.org/10.1016/j.wasman.2018.04.038

Shen, G., Li, Z., Hong, T., Ru, X., Wang, K., Gu, Y., Han, J., & Guo, Y. (2023). The status of the global food waste mitigation policies: Experience and inspiration for China. *Environment, Development and Sustainability, 26*(4), 8329–8357. https://doi.org/10.1007/s10668-023-03132-0

Simões, J., Pinto-Varela, T., Gaspar De Matos, M., & Carvalho, A. (2023). What influences consumer food waste in urban households? Guidelines for communication strategies in Portugal. *Journal of Cleaner Production, 429*, 139577. https://doi.org/10.1016/j.jclepro.2023.139577

STREFOWA. (2019a). *Appreciate your food and #reducefoodwaste—Guideline for Consumers 2019*. www.interreg-central.eu/STREFOWA

STREFOWA. (2019b). *Efficient treatment of food waste—Guideline for Waste Management sector 2019*. www.interreg-central.eu/STREFOWA

STREFOWA. (2019c). *Improve your daily business and #reducefoodwaste—Guideline for Retail sector 2019*. www.interreg-central.eu/STREFOWA

STREFOWA. (2019d). *Improve your loss ratio and #reducefoodwaste—Guideline for Primary Production sector 2019*. www.interreg-central.eu/STREFOWA

Talekar, S., Ekanayake, K., Holland, B., & Barrow, C. (2023). Food waste biorefinery towards circular economy in Australia. *Bioresource Technology, 388*, 129761. https://doi.org/10.1016/j.biortech.2023.129761

The World Bank. (2024). *GDP per capita (current US$)—European Union, Romania* [dataset]. https://data.worldbank.org/indicator/NY.GDP.PCAP.CD?locations=EU-RO.

Thorsen, M., Croad, T., Vincent, T., & Mirosa, M. (2022). Critical success factors for food waste reduction. *Cleaner Waste Systems, 3*, 100059. https://doi.org/10.1016/j.clwas.2022.100059

Thyberg, K. L., & Tonjes, D. J. (2016). Drivers of food waste and their implications for sustainable policy development. *Resources, Conservation and Recycling, 106*, 110–123. https://doi.org/10.1016/j.resconrec.2015.11.016

Tonini, D., Albizzati, P. F., & Astrup, T. F. (2018). Environmental impacts of food waste: Learnings and challenges from a case study on UK. *Waste Management, 76*, 744–766. https://doi.org/10.1016/j.wasman.2018.03.032

United Nations Environment Programme. (2021). *Food Waste Index Report 2021*. www.unep.org/resources/report/unep-food-waste-index-report-2021

Veselá, L., Králiková, A., & Kubíčková, L. (2023). From the shopping basket to the landfill: Drivers of consumer food waste behaviour. *Waste Management, 169*, 157–166. https://doi.org/10.1016/j.wasman.2023.07.002

Vizzoto, F., Testa, F., & Iraldo, F. (2021). Strategies to reduce food waste in the foodservices sector: A systematic review. *International Journal of Hospitality Management, 95*, 102933. https://doi.org/10.1016/j.ijhm.2021.102933

Zhang, Q., Dhir, A., & Kaur, P. (2022). Circular economy and the food sector: A systematic literature review. *Sustainable Production and Consumption, 32*, 655–668. https://doi.org/10.1016/j.spc.2022.05.010

Index

I. Ajtai et al. (eds.), *Guidelines for Food Waste Prevention and Reduction: Insights from the Fruit and Vegetable Sector*,
https://doi.org/10.1007/978-3-032-14506-2

MIX
Papier aus verantwortungsvollen Quellen
Paper from responsible sources
FSC® C105338
FSC
www.fsc.org

If you have any concerns about our products,
you can contact us on
ProductSafety@springernature.com

In case Publisher is established outside the EU,
the EU authorized representative is:
Springer Nature Customer Service Center GmbH
Europaplatz 3, 69115 Heidelberg, Germany

Printed by Libri Plureos GmbH
in Hamburg, Germany